# Statistics
## in theory and practice

# Statistics

## in theory and practice

**L R Connor**
MSc (Econ), FSS
Farr Medallist

and

**A J H Morrell**
MA, FIS, FSS
Formerly Principal Lecturer in Statistics,
Polytechnic of North London

Seventh edition by A J H Morrell

Pitman

PITMAN PUBLISHING LIMITED
39 Parker Street, London WC2B 5PB

*Associated Companies*
Copp Clark Ltd, Toronto
Fearon-Pitman Publishers Inc, Belmont, California
Pitman Publishing Co. SA (Pty) Ltd, Johannesburg
Pitman Publishing New Zealand Ltd, Wellington
Pitman Publishing Pty Ltd, Melbourne

Seventh edition 1977

Text set in 11/12 pt Photon Imprint, printed by photolithography,
and bound in Great Britain at The Pitman Press, Bath

ISBN 0 273 01109 X

# Contents

| | Preface | vii |
|---|---|---|
| 1 | Introductory | 1 |
| 2 | Statistical data | 4 |
| 3 | Routine statistics | 11 |
| 4 | Censuses and surveys | 21 |
| 5 | Statistical tables | 30 |
| 6 | Accuracy and error | 39 |
| 7 | Charts | 47 |
| 8 | Analysis of time series | 62 |
| 9 | Frequency distributions | 78 |
| 10 | Averages | 93 |
| 11 | Dispersion and skewness | 107 |
| 12 | Probability and some special distributions | 124 |
| 13 | Sampling | 142 |
| 14 | Regression and correlation | 165 |
| 15 | Comparing frequencies with expectations | 182 |
| 16 | Some industrial applications of statistics | 195 |
| 17 | Index numbers | 210 |
| 18 | Introduction to published statistics | 231 |
| 19 | Demographic statistics | 236 |
| 20 | Prices | 250 |
| 21 | Labour statistics | 257 |
| 22 | Production | 264 |
| 23 | Internal trade | 271 |
| 24 | Overseas trade and Balance of Payments | 277 |
| 25 | National income and expenditure | 282 |

v

**Appendices**

1  The use of computers for statistical work                    295
2  Suggestions for general reading                              298

   Answers to Exercises                                         300

   Index                                                        307

# Preface to the seventh edition

Having spent the last few years of my working life in trying to train statisticians, I have become increasingly aware of certain omissions from this book. It is always difficult to judge how far to go in an elementary textbook, but I decided that something must be said about chi-squared, distribution-free tests and probability paper. In addition, the later chapters on published statistics have been, as far as possible, brought up to date. The number of worked examples has been increased, particularly in Chapter 13. Finally, the Appendix on Data-Processing Techniques has been replaced by one on the use of computers in statistical work, with less technical detail than previous appendices.

I considered for some time whether to add a few statistical tables, such as are found in most books on statistics, but decided against it, partly on grounds of economy and partly because there are several easily accessible (and modestly priced) books of tables, such as those published by the Institute of Statisticians.

As in previous editions, I wish to thank the Institute of Cost and Management Accountants, the Institute of Certified Accountants, the Institute of Transport, and especially the Institute of Statisticians for permission to reproduce examination questions; the Controller of HM Stationery Office for permission to reproduce copyright matter; my colleague Frank Blewett for the new Appendix 1; and Mrs Josephine Parkinson for typing the new material.

A J H MORRELL

July 1977

**Abbreviations**

The following abbreviations are used to indicate the sources of examination questions reproduced as exercises at the end of suitable chapters:

ACA     *Association of Certified Accountants*
ICMA    *Institute of Cost and Management Accountants*
IoS     *Institute of Statisticians*
IoT     *Institute of Transport*

# 1 Introductory

The word 'Statistics' originally referred to collections of facts, not necessarily numerical, about the State, or the people who composed the State. According to modern usage it refers to collections of numerical facts or estimates and any restriction to affairs of state is obsolete. It may be construed either as singular or as plural. Statistics (plural) are the figures themselves, suitably classified and tabulated, together with any secondary statistics such as percentages or averages derived from them; it is in this sense that the public usually think of statistics. Statistics (singular) is the study, better described as Statistical Methods, which deals with the collection, analysis and interpretation of the figures.

Statistical Methods may be broadly divided into Descriptive Statistics and Mathematical Statistics. Descriptive Statistics deals with the compilation and presentation of data as actually recorded, generally in the course of public or private business, not for the purpose of refined analysis but simply in order to provide concise information on which decisions can be taken. Mathematical Statistics is based on the theory of probability and attempts to draw precise general conclusions from the data; it may also help to decide how the data can be obtained most efficiently and economically. In recent years there have been great advances in the design of experiments in biology, agriculture and industry with the aid of statistical methods. In economic and commercial problems, however, it is difficult to conduct experiments, and the scope for refined methods and mathematical techniques is severely limited.

Just as there are different kinds of statistics, so there are different kinds of statisticians. Indeed, the subject of statistics is now so wide that it is impossible for one man to become expert in all branches. The industrial statistician may know little about economic statistics, and the economic statistician may know little about the mathematical techniques used in scientific and industrial research. The present tendency, however, is to require some knowledge of the whole field including operational research and the elements of computers.

## The purpose of statistics

It cannot be too strongly stressed that the purpose of statistics is not simply to satisfy curiosity but to provide estimates or comparisons on which sound decisions can be made. Thus, if a manager is told that Factory *A* produced 500 tons of a certain product in a week, he must have some standard of comparison to tell him whether that is satisfactory. He may judge that the capacity of the factory is 650 tons a week; or he may be given a series of figures showing that it normally produces about 600 tons a week; or again, he may be told that Factory *B*, with half as many workers but otherwise comparable with Factory *A*, produced 350 tons in the same week. Such comparisons provide information on which action can be taken.

Looking further ahead, estimates based on present knowledge are essential to government and industry alike when they are formulating plans for the future. A local authority planning to build new schools must estimate the numbers of children likely to require admission during the next few years. A manufacturing firm deciding whether to build a new plant, or launch an export drive or an advertising campaign, requires estimates of future demand as well as knowledge of current trade and production statistics. Such estimates may be based partly on statistics of population, incomes, regional employment etc., and partly on a mathematical analysis of past records projecting into the future. They must, of course, be used with discretion, as forecasts may be falsified by political events, economic crises, new inventions and other unforeseen circumstances; nevertheless, they must be made, and as wisely as possible.

It has been said that every statistical table or chart should be preceded by the question 'What for?' and followed by the question 'So what?' This advice need not be taken too literally, but the idea behind it is sound. There is often a temptation to compile statistics for their own sake, which must be sternly resisted. *The purpose of statistics is to enable the recipients to take decisions.* The decision may be negative: that does not matter. If the statistics show that no action is required, that does not mean that they were wasted: they may save time that would otherwise be spent in argument.

It must always be remembered that statistics deal only with

measurable aspects of things and therefore can seldom give the complete solution to a problem; they provide a basis for judgment but not the whole judgment. Moreover, statistics can be, and often are, used, consciously or otherwise, to justify decisions reached on other grounds. It is therefore important that the limitations of statistics should be understood both by those who compile them and by those who use them. This warning will be repeated from time to time and in various forms throughout this book.

# 2 Statistical data

## Formulating the problem

Before initiating a statistical return or undertaking a statistical inquiry, the statistician should try to ascertain exactly what is required, and for what purpose. His duty is not only to compile statistics but to interpret them, and even, on occasion, to say that the question cannot be answered by statistics. He should therefore know what use is likely to be made of the results and, if necessary, how soon and how accurately they are required. The last two requirements may not be compatible.

It is important to ask the right questions in the first place and then to obtain the right answers. No sophisticated techniques can overcome defects in the basic data. As the computer experts say, 'Garbage in, garbage out'. The following points should be considered at the outset:

1. Can the required information be given in numerical terms, or can a good substitute for it be found? For example, ability or intelligence cannot be measured directly, but only by results, test scores or examination marks. The incidence of drunkenness can be measured only by numbers of convictions, which will depend partly on the strength or efficiency of the police.

2. What is the precise definition of the object to be measured? In a wage inquiry, for instance, should the data be wage rates or actual earnings, should allowance be made for overtime and bonuses, receipts in kind (e.g. free travel for railway workers), sick pay and holidays with pay, and should earnings be gross or net after PAYE, superannuation contributions etc.—always assuming that these alternatives are practicable? (See 1 above.)

3. What should be the field of the inquiry? As will be seen later (Chapter 22), the Index of Production includes extractive and manufacturing industry but excludes transport, distribution and ser-

vices because of the difficulty of finding suitable units of output.

4. Is there any information already available from routine statistics or published sources that will either obviate or supplement further inquiry?

The next question is how to collect the data. Often they can be obtained by direct observation, by counting or measuring. Examples are traffic censuses, times recorded by stop-watch, temperature readings and results of laboratory experiments. In fact, individual items of data are commonly called 'observations'. Sometimes, however, it is necessary to ask people for the information, either by interview or by some form of postal inquiry. This method will be considered further in Chapter 4.

Data may be divided into Primary and Secondary Data, although authorities differ as to the precise definition of these terms. Broadly speaking, primary data are the raw material, the figures collected at first hand, while secondary data are statistics taken from someone else after being worked up to some extent. The distinction is by no means clear-cut.

Another way is to classify data by source. They may be Internal or External according to whether they are collected by one's own organization or taken from published statistics or some other outside source. Internal statistics may again be divided into statistical records collected as a matter of routine, and special data obtained by *ad hoc* inquiries to answer specific questions. These two types of internal statistics will be considered in the following two chapters. The latter part of this book deals chiefly with published statistics, but the reader may find it helpful to glance now at Chapter 18 and read it more carefully later.

## Some common pitfalls

Statistics, and especially other people's statistics, are full of pitfalls for the user, and even official statistics must be examined critically. Typical sources of difficulty are:

### 1. *Changes in classification*

For example, up to June 1948, numbers registered as unemployed in the United Kingdom relate only to persons insured under the Unemployment Insurance Acts. From July 1948 onwards they include both insured and uninsured persons. See also the remarks on changes in the Standard Industrial Classification in Chapter 21.

### 2. *Changes in geographical and administrative areas*

As a result of the local government reorganization in 1974, County Reports on the Census of Population had to be produced on both the old basis and the new.

### 3. *Differences of definition*

Terms such as 'incomes', 'profits', 'output' and 'numbers employed' may be defined in various ways, and this frequently causes confusion. The student may like to try his hand at defining 'chemicals' and 'chemical industry'.

A certain factory in a large company found that, on average, its employees had about 40 per cent more sickness absence than those in the company as a whole. It turned out that they were including weekends in 'days lost' instead of taking working days only.

### 4. *Incompleteness of the information provided*

Thus, the annual inquiry by the Department of Employment into earnings and hours of work omits certain important industries such as coalmining and agriculture. (See Chapter 21.)

It is safe to say that statistics are hardly ever strictly comparable over a long period of years. The statistician must therefore take pains to ensure that he understands any figures he uses, and is aware of their defects and limitations. Equally, of course, he must be prepared to admit the limitations of his own statistics.

## Classification

Classification is the process of arranging things, either in fact or in the mind, in classes, each of which possesses some common property or properties relevant to the matter in hand. Thus, in a medical investigation, the incidence of a disease may very well vary between different sexes, different age-groups, different localities and different periods of time. Each of these factors should therefore be recorded for every case of the disease, so that the effect of each one may be evaluated and useful comparisons made between the various classes.

The above illustration exemplifies another important distinction between types of data, i.e.:

1. *Qualitative*, as when people or things are sorted in groups, each possessing some attribute that cannot be expressed numerically.

People may, for example, be classified male or female, married or single, or according to their occupation, eye colour, religion etc.

2. *Quantitative*, as when items vary in respect of some characteristic that can be measured or counted. Such a characteristic is called a *variable* or *variate*. Examples are age, height, income, examination marks, number of children.

The distinction between qualitative and quantitative factors, or attributes and variables, is important because variables lend themselves to mathematical treatment in the form of averages etc., while attributes do not.

## Homogeneity

The finer the classification, the less the chance of including unlike objects under the same head, and the better the chances of being able to compare like with like. For example, it would be misleading to compare the output per employee of two factories if one used more power per head, or more capital equipment per head, than the other. Furthermore, items should be so classified that the totals of the classes make sense. It would be absurd to add 400 motor-cars to 200 prams and call the result 600 vehicles.

It sometimes happens, however, that the variety in the items to be classified is so great that perfect homogeneity within each class can be achieved only by an elaborate and unmanageable classification. Iron and steel production is normally measured in tons, but there is no limit to the variations in shape, size etc., of iron and steel products, and they differ tremendously in cost per ton and the amount of work required per ton of metal. Non-ferrous metals and alloys are more complicated still: to compress statistical tables for such products into a reasonable compass it is essential to group together items that are not strictly comparable. The same remark applies to articles like drugs, paints and toys; it may even be necessary to express the totals of such items in terms of value only, not of quantity.

## Units

A continuous variable can be expressed only in terms of some unit: a distance in terms of yards, metres, miles etc., a weight in terms of pounds, tons etc., and so on. These are simple units, but it is often necessary to use composite units such as man-hours, foot-pounds, and feet per second or miles per hour. Some composite units involve three or more simple units. Thus, acceleration can be measured in feet or centimetres per second per second, abbreviated for convenience to ft/sec$^2$ or cm/sec$^2$. Again, power is measured in foot-pounds per minute or per second.

Care must be taken to ensure that units are, as far as possible, uniform and homogeneous. Thus, many acids and other chemicals are sold in solutions of varying strength, and it would be misleading to give their gross weights and add them together. The normal procedure is to express them in terms of 100 per cent acid or some other standard strength. Nitrogenous fertilizers are commonly expressed in terms of nitrogen content, i.e. as so many tons of nitrogen.

Units should be clearly defined and free from ambiguity. For example, in dealing with American statistics it must be remembered that the US gallon (but not the Canadian gallon) is five-sixths of the Imperial gallon, and that the US and Canadian hundredweight means 100 lb, not 112 lb. Tables headed 'tons' or 'thousands of tons' may still refer to long tons (2240 lb) as in Great Britain, short tons (2000 lb) as in America, or metric tonnes as on the Continent.

Units must be appropriate for their purpose, e.g. the labour force of an industry should be given in terms of employees, workers, 'productive' workers etc., whichever is most suitable in the context.

Finally, units should be stable. The outstanding example of an unstable unit is a unit of money, since it constantly diminishes in real value. One remedy is to express values in terms of the £ (say) of some standard year by means of a price index (see Chapter 17).

## Derivative data

A Statistical Derivative is a quantity formed by combining in some way two or more of the original items; the obvious example is a total. This section will be confined to simple derivatives such as ratios and percentages. More complex functions such as measures of dispersion will be considered in later chapters.

There are two ways of comparing two quantities of the same kind, say $x$ and $y$ units. One may be subtracted from, or divided by, the other according to whether it is the actual or the relative difference that matters. The difference $x - y$ is positive or negative according to whether $x$ is algebraically greater or less than $y$, but if the absolute difference is required, it is denoted by $x \sim y$ or $|x - y|$ and called the *modulus* of $x - y$. Thus $|3 - 5| = 2$.

The fraction $x/y$ is often expressed as a ratio $x:y$. Clearly the ratio is unchanged if $x$ and $y$ are both multiplied or divided by the same quantity, e.g. $15:5 = 75:25 = 3:1 = 1:1/3$. In words, '15 is to 5 as 75 to 25, or as 3 to 1 etc.'.

It is frequently necessary to compare two quantities of different kinds. Thus, a town covering 2450 acres and containing a population of 76,300 persons has a population density of 76,300/2450, or 31·1 persons per acre. The student may think of many other examples, such as speeds, volumes, densities and coefficients of expansion. Such averages may of course conceal great variations within the town, population etc.

Percentages must be used with care. Often it is not clear what they refer to, particularly when percentages of percentages are introduced. For example, a certain company was stated to have reduced its dividend 'by 100 per cent'. Further reading showed, not that the company had 'passed' its dividend as might be supposed, but that it had reduced the dividend from 200 per cent to 100 per cent. Such ambiguous statements should be avoided.

## Rates

When the numerator of a fraction or the first term of a ratio refers to the number of events or cases occurring during a certain period of time, the fraction or ratio becomes a rate, and is frequently expressed as so many per 1000. Thus the death rate is the number of deaths per 1000 of the population during the year in question. The death rate for a quarter is generally expressed as an *equivalent annual rate*, the number of deaths being multiplied by 4 and divided by the population, and similarly for any other period.

Care should be taken that the numerator and denominator are truly comparable, that the denominator represents as accurately as possible the number 'exposed to risk', as the actuaries say. A marriage rate, for example, should be calculated on the number of single and widowed persons of marriageable age, and a legitimate birth rate on the number of married women of reproductive age. (See Chapter 19.) Similar considerations apply to averages of the kind considered in the previous section. Thus, the output of a factory can be divided by the total number of employees, the total number of 'workers', or the total number of 'productive workers', and so on.

When, as generally happens, the number exposed to risk varies over the period, it is customary to take an average. On this basis, the accident rate in a factory might be computed by dividing the number of accidents occurring during the year by the average number of workers on fifty-two weekly payrolls. For a population, the mid-year figure is generally taken as the denominator.

Another important point to remember is that an overall rate depends very largely on the structure of the population. Death and sickness rates, for instance, vary greatly according to sex and age, and changes in the crude death rate, covering the whole population, may be due largely to the varying age and sex structure of the population. This point will be further discussed in Chapter 19.

# 3 Routine statistics

Much has been written both about market research and about published statistics, but the routine statistics compiled regularly for purposes of day-to-day administration have been strangely neglected. This is largely due to the enormous variety of such statistics and of the organizations collecting them, which makes it extremely difficult for any one person to speak with authority on the subject. However, the attempt must be made. It will be convenient to consider the statistics required by a large manufacturing firm, but much of what follows will apply equally to retail firms, local authorities, transport organizations etc.

Routine statistics are essential to the efficient management of all but the smallest organizations. Many of them, such as those that emanate from the company's accounts, are necessary to comply with the requirements of the law or of government departments. Others, such as statistics of sales and costs, are generally needed to ensure that a firm is run profitably, and others, such as statistics of stocks and production, to ensure that it runs smoothly.

Apart from financial and cost statistics, which will probably be left to the accountants, the statistics maintained by an industrial organization can broadly be classified as Sales, Production, Supply (i.e. purchases and stocks) and Personnel. Each of these groups will be dealt with in turn, but certain general considerations applicable to all of them must first be discussed.

## Basic records

Routine statistics are generally derived from (i) permanent records, e.g. of staff or customers, and (ii) current documents which are usually short-lived, e.g. invoices or supply indents. In either case, much unnecessary work can be avoided if the records are carefully designed at their inception.

In a large organization the figures initially compiled from the basic records may have to be submitted to a central department for incorporation with others. For this purpose a form of return is normally provided. This also should be designed with care to ensure that it is easy to complete correctly and to interpret when received. The form taken by these records and returns will depend on such factors as their ultimate use, the unit period covered by each return, the amount of detail and the degree of accuracy required, and the permissible delay.

Permanent records, particularly if they are also the source of statistical data, should normally be on cards or in loose-leaf form, so that it is easy to insert new ones or to sort them into any classification which may be required. If it is not desirable to sort the cards (because they are subject to constant reference) or if the amount of information to be recorded necessitates a large card, the 'visible index' type of card is to be recommended. These cards are filed, usually in trays contained in cabinets, so that one line of each card is visible when a tray is pulled out. This is known as the visible strip, and essential information can be entered either in full or in code, or coloured signals can be attached to indicate certain particulars of the customer or employee to whom the card relates. Thus, any customer whose turnover with the company exceeds £5000 per annum might be indicated by a red signal; at any time it is easy to count up the number of customers in this category by pulling out each tray in turn and simply counting the number of red signals. If cards are frequently removed from the trays, separate visible strips may be used which remain in position when the cards are extracted.

When current documents form the source of routine statistics it is generally necessary to transfer the required information to a summary sheet. It is easy to record far too much information for this purpose. For example, a Sales Manager might require up-to-date daily statistics of the total volume of each product ordered. This would be obtained by recording each order as it was received, and it would be sufficient to enter the quantity ordered in the column appropriate to the product. Details such as the customer's name, the order number and the packages involved might be entirely irrelevant and should, therefore, be rigidly excluded from the record.

A record card or return should be so designed that it is straightforward to complete and the descriptions and instructions used are unambiguous. If, for lack of space, it is not possible to make it fully self-explanatory, then precise instructions must be given to those responsible for completing it, and a check should be made soon after a new record or return is instituted to make sure that it is being completed correctly. Thus, when a statement of absence is required, it should be made clear whether this is to include very short periods of absence, such as lateness, or to be confined to periods of not less than a whole day or shift; similarly, the expression 'days lost' without further explanation might mean only working days, or it might include weekends and holidays. When calculations are required, e.g. percentages, the totals concerned should be fully defined, and it is useful also to indicate the degree of accuracy which is required, otherwise some returns will show the percentages carried to numbers of decimal places which are meaningless.

Routine statistics usually relate to the past; their significance lies in the extent to which they affect actions or decisions in the future. It is therefore important, in designing a system of routine statistics, to be clear about their ultimate purpose, for it must be possible to select the significant information and present that alone on the final statement. For example, a manager may require sales statistics for the purpose of detecting changes in the pattern of sales, and he is less interested in the figures which accord with expectations. In this case actual sales can be compared with forecast sales, and only those which differ by more than a specified amount need be presented to the manager. He knows then that every figure in the statement is of some importance, and he is saved the task of ploughing through a mass of figures on which no action is required.

## Coding

It is often convenient to code the information contained on a card or form, using letters or numbers to represent the various classifications. For example, staff functions may be coded as follows:

*Commercial:*

| | |
|---|---|
| Administration | 00 |
| Distribution | 01 |
| Purchasing and Supply | 02 |
| Sales | 03 |
| Transport | 04 |

*Production:*

| | |
|---|---|
| Works Management | 10 |
| Maintenance | 11 |
| Planning and Progress | 12 |
| Supervision | 13 |
| Instruments | 14 |

and so on. Instead of writing 'Distribution' the clerks simply write '01' in the appropriate space: this saves time and space and makes it easier to sort the cards and tabulate the information. The advantage of a double-digit code when, as in the example quoted, the classification consists of a number of main headings and a more detailed subdivision, is that a tabulation may be either general or detailed according to its purpose; for the general one it is sufficient to sort on the first digit, but for the detailed one it is necessary to sort on both digits. Coding is usually necessary when punched cards or computers are used.

### Punched cards

Cards into which holes are punched to denote the required information by their position are of two kinds: those used in connection with computers or with electrical or mechanical sorting and tabulating machines, and those which are sorted by hand with the assistance of special devices. With the latter type of punched card, holes are provided either round the edge or in part of the body of the card, and each hole represents a code number; when the information (which may also be written on the card) has been coded, the appropriate holes are elongated by a special instrument and the card is filed. When it is desired to select the cards relating to a particular code number, a needle is passed through the appropriate hole, through the pack of cards, and those which have been punched in this position either fall out or protrude, according to the particular type used.

Such cards are useful where it is occasionally necessary to extract statistical information but the job does not justify expensive equipment. For example, it might be convenient in a moderately large organization to record particulars of each accident on a punched card and periodically to examine all accidents due to failure to observe safety regulations, or those which resulted in some particular injury such as a broken leg.

Punched-card machinery is quite expensive, and a substantial volume of work is required to justify an installation. It is probably most economical when the punched cards recording the basic data can be used for a variety of purposes, e.g. preparing invoices on the tabulating machine as well as providing sales statistics. It is very valuable also where a number of classifications are required from the same basic data. For example, from a set of punched cards each representing one order, it would be possible to produce all kinds of tabulations for various purposes, e.g. quantity and value of each product purchased by each customer, customers purchasing each product, numbers of each type of package used, quantity of each product consigned to each town or county, and so on. Care should be taken, however, to ensure that the mere possession of a machine capable of producing all this information does not in itself lead to a demand for unnecessary tabulations.

**Fig. 1**  Punched card bearing personnel data

Fig. 1 shows a typical punched card that might be used by a personnel department. The holes punched denote the employee's name, address, date of birth, salary, date of joining the company, job level, department and nature of skills or qualifications.

## Sales statistics

Sales statistics merit prior consideration, for it is the demand for a company's products that sets in motion the whole complex machinery of production. Sales statistics, which consist essentially of the quantities of products supplied, may be compiled either from the orders as they are received or from the internal records subsequently prepared, e.g. copy invoices. Comparison with finished stocks and the production programme will show whether all is well or action is necessary to stimulate demand, e.g. by intensified advertising, or even to reduce it by a quota system, and whether any change in the production programme is needed.

Sales statistics in detail may show the sales to each customer (and indicate the attention to be given by representatives or the credit terms to be allowed), the sizes and packages required (to assist planning in the Packaging Department), and the volume of goods dispatched to various parts of the country (to determine the most economical method of dispatch to various places). They may also provide information on the ultimate use of the company's products by various industries and contribute to Government statistics on this subject. Sales statistics sometimes include details of inquiries received which have not resulted in orders, and 'competition intelligence', i.e. information on the proportion of a customer's trade held by a competitor.

Sales statistics are often required in two stages—'red-hot' daily figures (which need not be exact) for immediate sales action and production planning, and 'historical' exact figures for accounting purposes and long-term planning. The unit period may be a day for the first purpose and three months for the second; the previous day's figures may be required first thing each morning, but it may be immaterial if the quarterly return is not available until a month after the end of the period.

It is very easy to produce too much detail in sales statistics, and this can be very costly. Hence, it is essential that the exact use of the figures should be decided at the outset so that any irrelevant detail can be eliminated. It is also easy to produce exact statistics too often. A period such as a week, or even a month, may well be too short for any important trend in demand to reveal itself, and returns based on such a period may lead to waste of time in trying to find the reasons for fluctuations which are simply normal.

Sales forecasts are usually prepared periodically based on previous experience, market trends and surveys, and general information obtained by representatives from customers. The forecast may be quite simple in terms of total expected sales, or it may be detailed by localities or even down to individual large customers. The value of a forecast is that it provides a standard by which actual achievements may be judged, and it facilitates the detection of any failing or shortcoming in any part of the selling force.

Selling expenses may be analysed to ensure that the selling force is economically deployed and that advertising expense is directed into the most fruitful channels. There is some difficulty in allocating selling expenses, particularly when a representative is dealing with several products, and when an attempt is being made to build up a new market; some approximation may be necessary and the analysis should be treated with caution.

## Production statistics

Production statistics relate to the quantity and quality of output achieved, which is usually compared with the production plan or 'budget', and to such matters as machine-running time, individual workers or group performances, numbers and duration of breakdowns, and the quantity and value of spoiled material. This information is generally obtainable from workshop orders which detail the work to be done, or from records used to allocate workers' time over their various jobs.

The purpose of production statistics is to enable the Production Manager to ensure that he is fulfilling the demands of the Sales staff and operating economically. This again illustrates the essential function of statistics—control. The foreman 'knows it all'—he knows his men, he knows the work going through, he knows when a machine breaks down, and he does not need routine statistics to tell him these things. The Manager, however, cannot know it all and must depend on the routine statistics which are supplied to him.

## Supply statistics

Supply statistics will show primarily the quantity and value of the stocks of materials, components and equipment held. By comparing changes in stock levels the Supply Manager can see whether his buying policy is economical, and he can act to prevent either over-stocking or under-stocking. He is concerned to avoid tying up too much money in stocks, but at the same time to avoid running out of some vital component and thereby possibly causing a plant to stop production. By examining the fluctuations in demand for various items and seeing how long suppliers take to deliver, he can fix stock levels at which an order should be placed and the minimum level at which special action should be taken to replenish stocks. The basic record is usually a stock card for each item, on which the receipts and issues are entered; this card often provides for periodical calculation of the value of stocks held which can be reconciled with the accounts.

More detailed analyses can assist the Supply Manager further. The comparison of transport charges will help to determine the most economic mode of delivery for the principal items of stock. Evaluation of stocks of certain small items may show that the administrative cost of maintaining detailed records for these is unjustified and may lead to a reduction of clerical work. Statistics of orders typed will help to show whether the right size of staff is employed and whether there are seasonal fluctuations in the work. In these ways the routine statistics can give the Supply Manager the control which he needs.

## Personnel statistics

Personnel statistics yield information about the staff and labour force employed, wages and salaries, hours worked, absence and turnover. Detailed records of the behaviour of a large group of employees can be a very useful source of information for guiding personnel policy but, in this field particularly, so many possible classifications can be made that there is a great need for careful selection if the final statements are not to be so complicated that they conceal the salient features. The basic employee record usually includes the essential details of the employee himself, i.e. name, address, sex, age and possibly whether married and number of dependants, particulars of his education, training and previous employment, and information about his present employment with the company, i.e. date engaged, occupation, results of any tests, and so on. Changes of occupation and status and other matters of major importance will be recorded on this card as time goes on. Particulars of his timekeeping and absence record may be included or these may be more conveniently compiled on a separate record card.

In personnel statistics one rule that must be observed is to keep figures for men separate from those for women. Apart from this, the classification is largely a matter of choice, but it should be possible to analyse employees at any time according to their grade or occupation, factory or department, and it is desirable to be able to sort them into age groups and possibly into length-of-service groups. Division into married and single is more important for women than for men, but it may be important for men if there is a pension scheme.

Two problems of great concern to most large organizations are absence and labour turnover (i.e. numbers leaving the employer's service). It is not easy for an individual firm to influence either of these factors, but they can be so wasteful that it is well worth trying to analyse their causes. Sickness should be analysed by sex, category of workers etc. and, if possible, by age group and diagnosis. A thorough analysis involves coding and, if possible, the use of punched cards or a computer. For diagnoses, there is a three-figure International Statistical Classification of Diseases and Injuries compiled by the World Health Organization.

Labour turnover can be divided into voluntary (i.e. those who leave of their own accord) and non-voluntary (i.e. resulting from retirement, deaths and dismissals). It is notable that voluntary turnover is generally much heavier in the first few months of employment than later on, and an apparent high turnover rate may in fact be due to the 'fringe' of workers (perhaps 10 per cent) who are constantly changing their jobs; this can be an expensive business for the employer because of the administrative and training costs involved in the engagement of a worker. It is difficult to ascertain the real reasons for workers leaving, but it may often be done by skilful interviewers if they are given sufficient time. Turnover can, therefore, be analysed by sex, married or single (at least for women), length of service, age, grade or type of work and, very broadly, by causes of leaving.

One thing that must be stressed here is the importance of being able to compare any figures of sickness or turnover with the corresponding numbers exposed to risk. For example, if the number of workers leaving with less than three months' service falls from 500 to 300 in successive quarters, it may only mean that fewer men were engaged in the one quarter than in the preceding one. No valid deduction can be drawn from the numbers leaving unless the average number of employees in the category chosen can be calculated for each period. It is this sort of point that must be borne in mind when returns are designed.

## Exercises

3.1   An organization wishes to improve its records of sickness and absenteeism (absence without adequate reason). It is proposed to set up a card-index of employees to summarize information on a yearly basis.

Design the format for an *employee record card* giving:

(i) basic information, e.g. name, age, sex, grade (ancillary, unskilled, skilled, clerical, managerial), salary, department and duration of service; and

(ii) details of all absence during the year including study leave, official leave, permitted time off, illness and unspecified absence.
(*IoS*)

3.2   A manufacturing company wishes to compare production rates at its three factories. Design a schedule on which to record the data relating to orders received, production begun and production completed for each factory. (*IoS*)

3.3   Accident and attendance records are collected in most factories and time lost due to accidents is often compared between factories.

Give *two* reasons why errors occur in the collection of these data within a factory, and *two* reasons why the comparisons between factories might be misleading unless there is close control over data collection and presentation. (*IoS*)

# 4 Censuses and surveys

The previous chapter dealt with routine statistics collected for the continuous information and guidance of management, often as a by-product of administration. The present chapter is concerned with inquiries designed to answer specific questions or to provide certain information once and for all.

A census is generally a complete count of a population (or shops, farms etc.), whereas a survey is usually conducted on a sample basis and aimed at obtaining more general or more detailed information. The distinction, however, is by no means clear-cut, as is shown by the 'Sample Census of Population' in 1966 (see Chapter 19). Moreover, a census is generally much more than a mere enumeration, and may be used to obtain information about such matters as housing and car ownership.

## Full-scale or sample inquiry?

With a fairly small and preferably intelligent population, there is little difficulty. When the Institute of Statisticians conducts a Salary Survey, it sends a questionnaire to all its members with every confidence that the great majority of them will complete it correctly and return it promptly. When a market research organization wants to conduct a nation-wide inquiry, embracing all types of people, that is another matter. Such an inquiry is only practicable on a sample basis.

The principle underlying sampling is that a set of objects taken at random from a larger group tends to reproduce the characteristics of that larger group: this is called the Law of Statistical Regularity. There are exceptions to this rule, and a certain amount of judgment must be exercised, especially when there are a few abnormally large items in the larger group. With erratic data, the accuracy of sampling can often be tested by comparing several samples. On the whole, the larger the sample the more closely will it tend to resemble the population from which it is taken; too small a sample would not give reliable results. It will be shown later in Chapter 13 how the precision of estimates obtained from a sample can often be estimated from the sample itself.

The Law of Statistical Regularity may be exemplified by assurance statistics. Actuaries base their life tables and calculations of premiums to be charged on masses of data obtained from past experience, extending over many years. Experience shows that mortality rates do not vary greatly from one year to another, although they certainly have fallen over a long period and are still falling. Furthermore, the experience of all companies in any one year is practically the same, so that although the records of any one company are only a sample, they are clearly representative of the population as a whole, with the proviso that a small section of the population is uninsurable.

The advantages of a sample are that it costs much less and takes less time than a full-scale inquiry, and that the results are known more quickly; also, that it is possible to ask more questions and obtain more information than would be possible in a full-scale census. If the inquiry is made periodically, sampling enables it to be made more frequently. On the other hand, a sample nearly always means some loss of accuracy, but even here there are exceptions; a sample survey carried out with due care by trained people may give better results than a full-scale census taken indifferently.

The decision whether to take a sample therefore depends mainly on the cost of the inquiry in time and money, the delay in obtaining the results if a full-scale census is chosen and the degree of accuracy required. If, for example, the information is already available on punched cards which can be quickly sorted and tabulated, a full census is generally advisable. This might apply, for example, to an investigation into employee turnover in a large firm.

### The problem of non-response

When it is necessary to ask people for information, and particularly when this is done by a postal inquiry, it often happens that a large proportion of those approached do not reply. It is then difficult to judge whether those who do are representative of the whole population. It may be that most of the replies come from people who have an axe to grind or feel strongly on a particular issue. If so, the results of the survey will clearly be biased.

This problem of non-response, or partial response, does not arise with observations on animals, plants or inanimate substances, but it is a very troublesome problem in market or opinion research. It can occur with full-scale as well as sample surveys. Thus, if all members of a university are sent a questionnaire about their sex life, or all employees of a firm are asked for their views on a matter of company policy, and only 20 per cent reply, that 20 per cent may contain a high proportion of cranks or misfits.

In order to encourage a good response, the questionnaire should be made as simple as possible, and accompanied by a good covering letter explaining clearly why the survey is being conducted and why it is important for the recipient to complete the questionnaire, and stressing that replies are strictly confidential. A pre-paid business envelope should be enclosed and a reminder card can be sent if necessary. The replies received after the reminder has been sent should be compared with those received earlier. If they differ, it can be assumed that those who have still not replied resemble the second batch rather than the first.

With interviews, the problem of non-response is less acute and different in character: some people will be away from home and others too ill or too deaf to be interviewed; some may be night workers and many will be persistently out; some will refuse to co-operate, not surprisingly, since door-to-door salesmen often begin by pretending that they are doing market research. However, most people will respond if approached intelligently.

## Mail inquiry or interview?

A postal inquiry is an easy and inexpensive way of obtaining information, and for routine business and administrative inquiries it is the standard method. It allows time for thought and consultation with others (which may or may not be a good thing) and avoids any embarrassment in answering questions of a personal nature. Its chief drawback is the high non-response rate already mentioned.

In market research and public-opinion surveys, where a high response rate is essential, personal interviews are usually best. They are generally conducted by trained 'field staff' armed with a standard list of questions. The interviewer must be honest about the reason for the survey, and while he or she may explain any questions that are not clear to the person interviewed, must be careful not to suggest the answers or to bias them in any way. *Interviewer bias* can arise through an interviewer misunderstanding a reply or marking a wrong code number on his schedule, and *response errors* may be caused through faulty memory, a wish to please or impress the interviewer (the 'halo effect'), or the fear that a truthful answer may result in something to one's own disadvantage. Wilfully misleading answers are probably rare, but may sometimes be provoked by the interviewer's manner or by too many questions.

### The questionnaire

The importance of a good questionnaire can hardly be exaggerated. Even in an interview, where the interviewer reads the questions and himself records the answers, it is essential that the questions should be clear, framed in simple language, free from ambiguity, in logical order, and such that the respondents can and will answer them honestly and intelligently. Leading questions ('Do you agree that . . .') and emotive words ('big monopolies') should be avoided. Bias can be caused just as much by loaded questions as by non-response.

It is a common practice to run a *pilot survey* before the main inquiry, primarily to test the questions. For this purpose, a relatively small sample is chosen, which need not be a representative cross-section of the population, since at this stage the actual answers are unimportant. A certain amount of information may emerge, such as an estimate of the likely non-response or differences between different groups of people, but the main object is to see whether the questions and instructions are foolproof, and the whole questionnaire well designed.

### Selecting the sample

In order to select the sample, it is usually necessary to have the whole population listed in some way. An employer will have a file, or at least a card, for each employee, arranged alphabetically or in some other logical order. A firm may have a mailing list of actual or potential customers. Such a list, or set of lists, maps or similar records, is called a *sampling frame*; the two most commonly used in Great Britain are the Electoral Register (for persons) and the Ratings Lists (for dwellings).

For a sample to be truly representative, it is essential that the sampling frame should be complete, up-to-date and adequate for the purpose. A public-opinion survey should not be taken from one particular section of the population, e.g. those whose names are in the telephone directory, or readers of a particular newspaper, unless it is made perfectly clear that the survey relates only to that section.

There are various ways of selecting the sample, such as

(*a*) Random sampling.
(*b*) Systematic sampling.
(*c*) Stratified sampling.
(*d*) Cluster sampling.
(*e*) Quota sampling.

There are a number of variations and combinations of these methods, but these are the principal ones, and they will be described briefly in turn.

## (a) Random sampling

Random sampling simply means choosing a sample in such a way that at each step all members of the population not already selected stand the same chance of being chosen. Thus a hand of cards dealt from a properly shuffled pack is a random sample; winning tickets in a lottery are chosen at random by taking numbers out of a revolving drum. By a simple extension of this idea, a random sample can be taken from a group or population by numbering all items of the group and picking numbers by means of a table of 'random numbers', i.e. numbers of (say) two or four digits compiled by some random process.

Note that 'random' does not mean 'haphazard', just thinking of a number. The method must be completely objective. If you doubt this, try asking a large class of students to write down a digit at random—any digit from 1 to 9. Probably about 30 per cent of them will put down '7'. There is a bias against 1 and 9, and possibly 5, because they do not 'look random', but no one really knows why 7 is so popular. Provided the sampling frame is sound, a random sample will be completely unbiased. This cannot be said of any other method.

## (b) Systematic sampling

This means taking the sample in some systematic manner, generally by taking items at regular intervals. Thus, in an agricultural experiment, a plot of land could be divided into strips of equal width, and every tenth strip selected. The same method can be used in sampling from a list, provided there is no systematic arrangement in the list, or that the order of the list has no bearing on the subject of the survey. A

10 per cent sample survey of sickness absence has been conducted in a large organization, taking every worker whose works number ended in a specified digit. Another possibility would be to take every person whose birthday occurred on a particular date in the month. Such samples are sometimes called *quasi-random*, because they are as good as random for the purpose in hand.

### (c)  Stratified sampling

This method makes use of prior knowledge of the population, and is the most commonly used type of sample design. The sampling frame is first divided into natural sections or *strata* (plural of *stratum* so, whatever you do, avoid talking about 'stratas'). This will probably be done already, if the frame is the Electoral Register or some other records held by local authorities scattered over the country. The strata may be regions, towns, streets, sexes, age-groups, social classes, occupations, and so on, depending on the nature of the inquiry. Within each region, a random sample of towns, rural districts etc. may be chosen. In fact, sampling may be done at several stages (multi-stage sampling). Finally, within each stratum or sub-stratum selected, a random sample of the required size is chosen.

There are many advantages in stratification. It ensures that each section of the population is adequately represented and that there is sufficient information about that section. It is less expensive than pure random sampling, because field staff can concentrate on a limited number of areas instead of having to visit one man in Skye or in the middle of Dartmoor. It is necessary, however, that the proportions of the population in the various strata should be known beforehand, at least fairly accurately.

### (d)  Cluster sampling

This is a variation of stratified sampling where, instead of taking a random sample within each sub-stratum selected, the whole of the sub-stratum is taken. It might, for example, be every child in a class or every member of a household. It is, in fact, a multi-stage sample with 100 per cent sampling (if that is not a contradiction in terms) at the last stage.

Cluster sampling allows a large sample to be taken for a given cost. It is useful if there is no comprehensive sampling frame for the whole population, or if there is much greater variation *within* clusters than *among* clusters.

## (e)  Quota sampling

This might be described as non-random stratified sampling. The total sample within each selected area is broken down according to sex, age group and social class, and the interviewer is given complete discretion as to how he chooses the individuals. This procedure has obvious dangers, as shown by the following incident. A survey was being conducted into betting habits. An interviewer arrived in London early one morning and decided to take some interviews before going to the office. What better than a queue of people waiting for a train, with nothing to do? Looking at the results afterwards he found they were hopelessly biased, and on making inquiries, discovered that the train was bound for Newmarket Races.

Quota sampling is cheap and easy, requires no sampling frame and no call-backs, and is very useful in certain types of inquiry, but it is biased towards people working out of doors and people with a good education. The judgment of social class is obviously very subjective. The main drawback, however, is that whereas it is possible to put estimates of reliability (see Chapter 13) on the results of a simple random or stratified sample, this is not possible with quota sampling.

## Public opinion polls

Most public opinion polls should be interpreted with caution, particularly those taken just before a General Election, which have sometimes given such misleading results as to bring these polls into disrepute. They are not in themselves forecasts: they only record voters' declared intentions at a particular time. Apart from the 'Don't knows', who may in the end sway the election, some will change their minds before the election, and many, for various reasons, will not vote at all. Some of these may be lulled into a false sense of security by the poll results themselves, and think that their votes are unnecessary. There may also be a campaign, in some quarters, to give deliberately false answers.

## Suggestions for further reading

Ilersic, A. R. *Statistics*. HFL (Publishers) Ltd.

Kish, L. *Survey Sampling*. London and New York: John Wiley.

Moser, C. A. & Kalton, G. *Survey Methods in Social Investigation*. London: Heinemann.

Yates, F. *Sampling Methods for Censuses and Surveys*. Griffin.

### Exercises

4.1  A large company wishes to gather information about the newspaper-reading habits of the general population, in order to find any correlation between reading habits and other characteristics such as age, sex, education etc. Design a questionnaire, to be used postally, to give such information as you think appropriate.  (*IoS*)

4.2  Write a brief explanation of the following terms used in market research

(*a*) sampling frame;
(*b*) interviewer bias;
(*c*) non-response rate;
(*d*) non-sampling errors.  (*ICMA*)

4.3  Statistical information may be obtained by the use of questionnaires. What steps should be taken to ensure that:

(*a*) the questionnaire is adequately designed; and
(*b*) that the method of collection (by post or direct interview) does not influence the results?  (*ACA*)

4.4  Before any large survey a pilot survey is usually carried out.

(*a*) Why is a pilot survey needed?
(*b*) What can it reveal?
(*c*) Who should carry it out?

Answer in note form.  (*IoS*)

4.5  A manufacturing company sent several interviewers out on to the streets with instructions to each to ask fifty people what they thought about the company's products. If this was all the instruction given to the interviewers, what sort of errors and bias would appear in the answers? Answer in note form, giving a brief example to illustrate each point.  (*IoS*)

4.6  Why do statisticians often use sampling techniques to obtain information? What are the purposes and procedures involved in random, systematic and stratified sampling?  (*ACA*)

4.7 Define and evaluate the sampling methods you would use to select a sample in the following surveys:

(*a*) to ascertain urgently the likely readership of a proposed weekly magazine devoted mainly to television and radio programme details, with articles on the programmes and artists;

(*b*) to provide information concerning the present health of former employees of a particular industry who have been incapacitated by a common disease. It has been established that 5000 persons were disabled in the period being considered and that a sample of 250 persons is sufficient;

(*c*) to assist a town council of a popular holiday resort, in connection with a proposed advertising campaign, with information of accommodation to be available during the coming season for visitors and tourists. (*ICMA*)

4.8 A market research agency was asked to interview as many women as possible among 1200 married women selected at random from a voting list for a district in a large town. The interviews were to be completed in a week and each woman was to be asked whether she had heard of a cake-mix called 'Fairymix'. If the reply was 'Yes', the respondent was to be asked if she had a packet in her larder.

The results were as follows

| | |
|---|---|
| Number interviewed | 550 |
| Number who knew of Fairymix | 100 |
| Number who had a packet | 23 |

Comment on the usefulness of this survey to the manufacturers of Fairymix. (*IoS*)

# 5 Statistical tables

When the statistician has collected his data in the form of cards, schedules or other documents, he may be faced with an amorphous mass of figures which must somehow be crystallized into a reasonably compact statement. Both the final product and the intermediate stages will probably consist largely, if not entirely, of statistical tables, since they provide a convenient way of summarizing the data in an orderly manner and of presenting the results concisely and intelligibly.

## Structure of tables

A table is a rectangular arrangement of figures, with headings describing the various rows and columns. The first column, containing the headings for the rows, is called the *stub*, and the headings of the other columns form the *captions*, while the figures constitute the *body* of the table.

**Table 1**  Mark-sheet of examination candidates and their marks

(An example of two-dimensional tabulation)

| No | Name | English | French | Mathe-matics | History | Total |
|----|------|---------|--------|--------------|---------|-------|
| 1 | Abbott, T. R. | 54 | 62 | 37 | 55 | 208 |
| 2 | Ackroyd, B. J. | 71 | 58 | 70 | 64 | 263 |
| . | . . . . | . | . | . | . | . |
| 113 | Young, C. | 67 | 48 | 82 | 63 | 260 |
| | *Total* | 6,392 | 6,205 | 5,977 | 5,649 | 24,223 |

The simplest form of table is a single column of figures with the stub, like a list of examination candidates and their marks in a single subject. The term 'table', however, generally conveys the idea of a set of figures arrayed in at least two dimensions, such as a mark-sheet giving the candidates' marks for a number of subjects, with totals, as in Table 1.

Notice that there are two marginal sets of totals, those for each candidate and those for each subject. The former are, of course, essential. The latter might be thought unnecessary, but they provide a useful check on the accuracy of the last column because the sum of the subject totals and that of the candidates' totals should agree in the same grand total. It is always desirable to make a table self-checking whenever possible.

Tabulation in three or more dimensions is not possible in the literal sense, but the same effect may be obtained by subdividing rows or columns or both. Table 2 is an example of tabulation in three dimensions, each of the main column headings being subdivided into Male, Female, and Total.

**Table 2**  Summary table showing numbers employed, total earnings and hours worked in Brown and Green Ltd in the week ended 5 March 1977

(An example of three-dimensional tabulation)

| Fac-tory | Nos employed | | | Total earnings | | | Man-hours worked | | |
|---|---|---|---|---|---|---|---|---|---|
| | M | F | T | M | F | T | M | F | T |
| | | | | £ | £ | £ | | | |
| A | 150 | 46 | 196 | 8,330 | 1,742 | 10,072 | 6,681 | 1,871 | 8,552 |
| B | 228 | 143 | 371 | 13,827 | 4,708 | 18,535 | 9,802 | 6,114 | 15,916 |
| C | 570 | 269 | 839 | 35,427 | 9,195 | 44,622 | 27,719 | 12,325 | 40,044 |
| D | 251 | 73 | 324 | 16,074 | 2,967 | 19,041 | 11,348 | 3,126 | 14,474 |
| Total | 1,199 | 531 | 1,730 | 73,658 | 18,612 | 92,270 | 55,550 | 23,436 | 78,986 |

In Table 3 the process has been carried a stage further. Again the main column headings are subdivided, this time into years, but the row headings for commodities are also subdivided, not only into more detailed commodities but in some cases into country of destination. Both kinds of sub-headings are shown slightly indented to the right—too slightly to make it clear which are sub-headings of what. The compilers have tried to compress the maximum of information into the page, thus suffering some loss of clarity. It is only fair to say that the trade accounts are no longer published in this form.

Too complicated classification and subdivision of headings make a

**Table 3**  Extract from UK trade accounts for January 1953
(An example of four-dimensional tabulation, now happily obsolete)
II—Exports (produce and manufactures of the United Kingdom)

| Class III<br><br>Articles wholly or mainly manufactured | | Quantities | | | Value | | |
|---|---|---|---|---|---|---|---|
| | | Month ended 31 January | | | Month ended 31 January | | |
| | | 1951 | 1952 | 1953 | 1951 | 1952 | 1953 |
| **GROUP F.—ELECTRICAL GOODS AND APPARATUS—** | | | | | | | |
| Electric cables, wires, strips and strands, insulated— | | | | | £ | £ | £ |
| Telegraph and telephone cables and wires— | | | | | | | |
| Submarine | Tons | 263 | 436 | 75 | 35,921 | 60,234 | 12,171 |
| Other— | | | | | | | |
| To Union of South Africa | ,, | 228 | 154 | 356 | 57,748 | 53,313 | 129,571 |
| To India | ,, | 40 | 67 | 59 | 11,595 | 25,393 | 39,210 |
| To Malaya | ,, | 204 | 25 | 68 | 42,670 | 10,467 | 18,259 |
| To Australia | ,, | 1,118 | 130 | 82 | 248,833 | 49,159 | 39,045 |
| To New Zealand | ,, | 315 | 470 | 247 | 77,728 | 129,574 | 104,162 |
| To Other Commonwealth Countries | ,, | 352 | 304 | 382 | 76,858 | 98,831 | 114,687 |
| To Irish Republic | ,, | 218 | 36 | 27 | 49,872 | 11,777 | 21,386 |
| To Argentine Republic | ,, | 98 | 19 | — | 32,010 | 9,767 | — |
| To Other Foreign Countries | ,, | 122 | 422 | 480 | 43,944 | 153,215 | 178,279 |
| *Total* | | 2,695 | 1,627 | 1,701 | 641,258 | 541,496 | 644,599 |
| Other descriptions— | | | | | | | |
| Paper insulated— | | | | | | | |
| To British West Africa | ,, | 127 | 67 | 31 | 26,259 | 15,015 | 9,383 |
| To Union of South Africa | ,, | 429 | 105 | 244 | 77,339 | 27,435 | 67,763 |
| To India | ,, | 417 | 622 | 1,469 | 75,647 | 136,074 | 382,802 |
| To Malaya | ,, | 127 | 122 | 392 | 22,876 | 30,606 | 104,725 |
| To Hong Kong | ,, | 18 | 63 | 71 | 3,401 | 16,253 | 16,878 |
| To Australia | ,, | 751 | 745 | 936 | 137,731 | 163,399 | 267,035 |
| To New Zealand | ,, | 208 | 58 | 79 | 38,172 | 16,012 | 22,930 |
| To Other Commonwealth Countries and the Irish Republic | ,, | 568 | 492 | 1,252 | 100,613 | 124,032 | 307,030 |
| To Norway | ,, | 34 | 32 | — | 6,793 | 7,620 | — |
| To Portugal | ,, | 30 | 18 | 11 | 5,329 | 3,748 | 2,715 |
| To Turkey | ,, | — | 147 | 181 | — | 32,733 | 43,374 |
| To Egypt | ,, | 4 | 122 | 17 | 730 | 30,359 | 5,459 |
| To Other Foreign Countries | ,, | 272 | 571 | 327 | 48,546 | 141,270 | 99,826 |
| *Total* | ,, | 2,985 | 3,164 | 5,010 | 543.436 | 744,556 | 1,329,920 |
| Rubber insulated— | | | | | | | |
| To Union of South Africa | ,, | 201 | 163 | 128 | 61,440 | 71,931 | 55,560 |
| To British East Africa | ,, | 70 | 97 | 26 | 26,856 | 39,127 | 13,358 |
| To India | ,, | 243 | 151 | 190 | 57,972 | 61,644 | 76,020 |
| To Pakistan | ,, | 95 | 91 | 9 | 30,359 | 53,306 | 3,158 |
| To Malaya | ,, | 57 | 150 | 98 | 15,863 | 71,692 | 35,870 |
| To Hong Kong | ,, | 150 | 60 | 61 | 41,531 | 26,420 | 25,235 |
| To Australia | ,, | 27 | 59 | 31 | 14,662 | 24,979 | 11,998 |
| To New Zealand | ,, | 199 | 233 | 67 | 54,161 | 116,034 | 28,759 |
| To Other Commonwealth Countries and the Irish Republic | ,, | 205 | 277 | 353 | 65,306 | 150,751 | 158,529 |
| To Foreign Countries | ,, | 238 | 300 | 263 | 90,268 | 145,583 | 129,689 |
| *Total* | ,, | 1,485 | 1,581 | 1,226 | 458,418 | 761,467 | 538,176 |

table difficult to read, and it is generally better to break it up into a number of separate tables or to condense the information.

## Stages in tabulation

The amount of tabulation necessary varies considerably according to the nature of the inquiry and of the data available. A typical case, involving the analysis of a number of schedules bearing information on several points, might contain the following stages:

1. Extracting the relevant items of information from the schedules and entering them on 'working sheets' designed to facilitate calculations.

2. Performing the necessary calculations with the aid of these working sheets.

3. Transferring the results of these calculations—totals etc.—to one or more summary sheets and subjecting them to further manipulation.

4. Compiling one or more summary tables embodying the results in all necessary detail, but omitting original and intermediate figures which have served their purpose. These tables will probably include derivative data such as averages and percentages. For example, a summary table derived from Table 2 might give average weekly earnings and hours worked in each factory and in the firm as a whole, together with percentage increases in average earnings since a given date.

## Working sheets

Working sheets are generally kept as a permanent record of the calculations performed on the data, both as a guide for future occasions and for reference in case figures are queried or further detail is required. They should, therefore, be drawn up carefully on good paper that will not tear easily. It is often worthwhile spending considerable time on designing a working sheet so that the calculations can be done as efficiently as possible. If many similar working sheets are required they should be stencilled or otherwise reproduced with all possible detail such as headings, footnotes, units, £ signs, and cross-references between items, saving the compiler both time and thought.

Plenty of room should be allowed for amendments and corrections. Sheets should be given proper titles, and columns their proper headings; continuation sheets should be numbered consecutively.

Units should be clearly shown and, if necessary, the source of the data should be stated. As mentioned earlier, whenever possible the calculations should be self-checking; a control column can be introduced if necessary to provide a check.

When a large number of items have to be counted, a useful device is to make a vertical stroke for each of the first four items and a stroke across them for the fifth, forming a gate. Thus, thirty-seven items would be recorded as follows:

This 'gate system' makes counting much easier and more accurate than a mere succession of strokes.

### Hints on the construction of tables

1. Having thought out what the table is to show, give it a clear and comprehensive title. Be particularly careful about dates and geographical regions: some official figures, for example, refer to the United Kingdom, others to Great Britain, and others to England and Wales; if the table covers the years, say, 1960–1976, this can be shown in the title, e.g.:

### UK PRODUCTION OF PIG IRON, 1960–1976

2. Consider the column and row headings carefully, and make sure they express exactly what is intended. Show totals and sub-totals clearly, if necessary by heavy type for the relevant headings and figures or by light and heavy rulings, and show sub-headings in the stub by judicious indenting. It is sometimes desirable to show totals at the top or in the first column of the table, particularly if only some of the component items are shown and they do not add up to the total.

3. State the units clearly, at the head of each column, or on the right-hand side of the stub (see Table 3) or, if all the figures in the table are in the same units, above the whole table. If figures are given to the nearest hundred, thousand, or million, the unit can be given as thousand tons, £ million etc. Much space and labour can be saved by working to three- or four-figure accuracy.

4. Make sure that comparisons shown in the table are valid. For example, output in periods of sometimes four, sometimes five weeks should be shown in weekly averages, and in presenting annual and

quarterly figures side by side, the quarterly figures can often be given as equivalent annual rates. Sickness rates for different groups of workers should not be shown for comparison without comment if the age-compositions of the groups are very different.

5. Give the source of your information at the foot of the table where necessary: nothing is more exasperating than to be unable to trace the source of a table. If it is a regular publication, give the date of it; figures may be revised in a later edition.

6. Any definitions or explanations that cannot be conveniently shown in the table should be given in footnotes or other accompanying notes, with numbers, asterisks or other symbols against the items or headings to which the notes refer. Any break in continuity through changes in classification etc., must be clearly shown.

7. Avoid complicated tables if you can but, if not, it may be helpful to number the rows or columns. The relations between them can be shown simply, as in the following example:

| *Gross Output* | *Raw Materials* | *Net Output* |
|:---:|:---:|:---:|
| (1) | (2) | (3) = (1) − (2) |

With a very wide table, repeat the row headings or numbers on the right-hand side.

8. With a long table, e.g. one involving a long list of countries, leave a space after every fifth item. With monthly data it may be preferable to give blocks of three or six months, thus:

1977: January
February
March

April
May
June

9. The modern tendency is to reduce vertical lines in typed or printed tables to a minimum, e.g. to put them between main headings but not between sub-headings in a table like Table 2. If figures are to be written by hand it is best to draw vertical lines to keep the columns straight.

Some of the above hints apply equally to working sheets. The list is by no means exhaustive, but the student who wishes to learn how to draft tables cannot do better than study the *Monthly Digest of Statistics* and other official publications.

### Exercises

5.1   Summarize the following information in, at most, two tables.
'The total volume of exports and re-exports of milk products from the United Kingdom for 1971 was 82,604 tons, of which 30,598 tons went to the Commonwealth. Africa received the biggest share with 12,290 tons which consisted of 11,611 tons of milk, 351 tons of cheese and the rest of butter. The West Indies received 465 tons of butter, together with 7225 tons of milk and 83 tons of cheese, and had the second highest total. South-east Asia's total of 3633 tons consisted largely of milk and only 7 tons of butter and 164 tons of cheese. Other Commonwealth countries accounted for the remaining 5740 tons of milk, 675 tons of cheese and 487 tons of butter. UK exports and re-exports of milk products for 1971 to foreign countries (i.e. outside the Commonwealth) included 46,716 tons of milk and 3229 tons of butter, the rest being taken in cheese. Total exports for 1971 were 12,933 tons more than in 1970. In the case of milk, the 1970 figure was exceeded by 10,188 tons. Butter, too, was up by about 150 per cent on the 1970 figure of 1852 tons but the figure for cheese showed a relatively small increase.'                                                       (*IoS*)


5.2   The following facts about three London boroughs are taken from the 1966 sample census of England and Wales:

There were 47,740 households with one car and 9900 with two or more cars in Bromley, 27,460 with one car and 3550 with two or more cars in Greenwich, 14,500 with one car and 910 with two or more cars in Tower Hamlets. The number of cars in the three boroughs was 68,740, 34,940 and 16,370 respectively.
An analysis of garaging arrangements showed that 10,970 cars were left at the roadside at night in Bromley, 13,800 in Greenwich, and 9760 in Tower Hamlets.
The total number of households in each borough were: Bromley 96,990, Greenwich 74,050 and Tower Hamlets 64,320 and the number of people in these households 289,760, 220,850 and 187,580 respectively. The area of each borough is 39,293, 11,724 and 4994 acres respectively.

Consider the foregoing and then:

(*a*) draw up tables to present a concise summary of the information;

(*b*) calculate the percentage of households in each borough without a car;

(*c*) calculate the percentage of cars in each borough left at the roadside at night;

(*d*) say how far the variation in (*c*) is related to population density.

(*IoT*)

5.3   In a market survey, housewives were interviewed to determine their preferences for, and use of, detergents. 100 housewives over 40 years of age were interviewed, and 150 under 40; 80 of the first group and 130 of the second group said that they used some kind of detergent. In answer to a second question, 'Do you use detergents for

(*a*) clothes washing,

(*b*) household purposes (e.g. washing floors etc.),

(*c*) other uses (e.g. car washing)?',

50 answered 'Yes' to part (*a*) from the over-40 group and 70 of the under-40s said they used detergents for clothes washing. In answer to parts (*b*) and (*c*), 70 and 5 respectively answered 'Yes' in the over-40 age group. The corresponding figures for the under-40s were 100 and 10.

Tabulate the data, adding any secondary statistics which may be helpful in illustrating the results of the inquiry.       (*IoS*)

5.4   Arrange the following information in a convenient tabular form, and fill in any gaps in your table:

The number of goods vehicles licensed in the United Kingdom increased by approximately $2\frac{1}{2}$ per cent, from 1,513,000 in 1964 to 1,551,000 in 1970. The number of goods vehicles of up to $1\frac{1}{2}$ tons (unladen weight) rose from 872,000 to 933,000 in this time, an increase of 61,000 or 7 per cent. The increase was most dramatic in the case of vehicles of over 8 tons, viz. 175 per cent, an increase of 35,000 on the 1964 figure of 20,000. There was also a marked rise in the number of goods vehicles of 5 to 8 tons, from 70,000 in 1964 to 132,000 in 1970, an increase of 89 per cent. The number of goods vehicles of intermediate weight actually fell in the period in question. For $1\frac{1}{2}$ to 3 ton vehicles the fall was from 291,000 to 197,000 or 32 per cent, while the number of 3 to 5 ton vehicles went down by 26,000 from 260,000 in 1964 to 234,000 in 1970.       (*IoS*)

5.5   Arrange the following information in a convenient tabular form:

Out of a total intake of 713 students to a college, 396 studied Letters, the remainder studied Science. The overall pass-rate for the students in their examinations was 73·4 per cent, the pass-rate being much higher for Letters (86·6 per cent) than for Science (56·8 per cent). The highest percentage of referred passes occurred among the Science students (26·2 per cent) with the students of Letters well below the total intake rate of 15·7 per cent. Only 6·1 per cent of the Letters students failed.

The average mark obtained by the total intake was 142·9 with the students of Letters and Science averaging 5·8 more and 7·3 less than this figure, respectively. The standard deviations of marks were 17·0 for Letters, 35·7 for Science and 27·7 for the total intake.   (*IoS*)

5.6   (*a*) Summarize the following information in an appropriate table:

'Of the 48 bulk carriers of 100,000 tons dead weight and above, built 1967–70 inclusive, three-quarters were built in Japan. Of the rest, half were built in Sweden and the remainder consisted of 3 built in Germany, 2 built in Northern Ireland (both of them for the Norwegians) and one built in England (for Britain). Of the ten carriers that fly the Norwegian flag, Sweden and Japan each built 3 and Germany and Northern Ireland shared the building of the rest. All the 28 carriers flying the Japanese, Liberian and Brazilian flags were built in Japan, only 2 being for Brazil, 11 for Liberia and the remainder for Japan herself. In the period referred to Britain had 2 carriers built in Sweden and 3 built in Japan. Germany and Sweden each built one carrier for their own use and Sweden had 2 more built in Japan.'

(*b*) List three outstanding features of the data.   (*IoS*)

# 6 Accuracy and error

Suppose it is desired to estimate the percentage by weight of nitrogen in a fertilizer mixture. Samples may be taken from different parts of the mixture on different days, sent to different laboratories, and analysed by different analysts using different methods. In the first place, there will be slight variations in the composition of the mixture from day to day and during the day, owing to changes in temperature, humidity and other factors. These variations are called sampling errors, but they are not errors in the usual sense of the word. In the second place, there will be errors of analysis, i.e. differences between laboratories, between methods used, and between individual analysts. Thirdly, there will be errors of measurement or observation.

There are thus several kinds of errors in experimental work, e.g. sampling errors, analytical errors and errors of observation, and they can never be completely eliminated. It should be noted that they are not necessarily mistakes. An error, in the statistical sense, is merely the difference between an estimate and the true or ideal value, which it may be impossible to determine exactly. It is these 'errors' due to the natural variability of observational data that create the need for statistical methods.

Different types of data naturally have different degrees of variability. In the physical sciences, variability is generally small, in biology and the social sciences it is relatively large, and in economics and commerce, where the statistician cannot experiment but must be content to observe, it is greater still, and the data themselves are subject to different kinds of errors from those met in experimental work. They may be broadly classified as errors of origin, errors of inadequacy and errors of manipulation. Errors of origin are due to defects in the data collected, e.g. false information caused by forgetfulness or misunderstanding on the part of the person supplying it (see page 24); errors of inadequacy arise when the items observed are too few to provide a representative sample of the population; and errors of manipulation are mistakes in counting or measuring, clerical errors etc., in fact any mistakes made in handling the data.

## Bias

If a quantity is such that its errors all tend to lie in the same direction, they are said to be biased. This does not mean that all errors are necessarily in the same direction, but there will be more in one direction than in the other, and the total of a number of items will almost certainly err on the side of the bias. For example, a table of incomes based on income-tax returns would show a downward bias because taxpayers are more likely to understate their incomes than to overstate them.

   If errors on both sides are equally likely, they will tend to neutralize one another and are said to be unbiased or compensating errors. Thus, if a number of large sums of money are rounded off to the nearest thousand pounds, the errors are unbiased, varying between −£500 and +£500.

## Accuracy and precision

By accuracy we mean the closeness of a figure to its 'true value'. Precision is not quite the same thing, although the two words are commonly used as synonymous. If a number of estimates, obtained by a biased method or with faulty apparatus, agree closely among themselves, they are precise but not accurate. If the rate of return on a proposed investment is given as 12·873 per cent, that figure is precise but may be far from accurate!

   Not only is absolute accuracy generally impossible, but it frequently happens that such precision as can be obtained is unnecessary; every schoolboy is taught that a good approximation to $\pi$ is $3\frac{1}{7}$. The accountant must, of course, balance his accounts exactly, if only to guard against fraud, but it does not matter to the Managing Director whether the week's turnover was £35,485·63 or £35,486·32, and it may be sufficient to round off the amount to £35,500 or even £35,000. A fair estimate produced quickly may be much more useful than a precise figure three months late.

   Very often all that is required is a rough estimate, and it is a waste of time to aim at needless precision. A slide rule will not normally give results correct to more than three significant figures, but it is an extremely useful device all the same. It is desirable, however, to show clearly what the precision of the figures presented really is, and this can be done in various ways. Generally it is implied by the figures or tables themselves. Thus, if sums of money are shown as £256,000, £73,000 and so on, or better still if a column is headed '£000', it is clear that they have been rounded off to the nearest thousand pounds.

Another method is to show the range within which the true figure is believed to lie. Thus a population might be stated as 267,000 ± 1500, meaning that the estimate of 267,000 might be as much as 1500 out either way. Another way would be to say that it lies between 265,500 and 268,500. Percentages are sometimes useful here: if a figure is believed to be correct within, say, 0·2 per cent, it can be shown as 463,000 ± 0·2 per cent.

In arithmetical computations, however, it is a mistake to approximate too soon, especially if the final result is the difference of two nearly equal numbers. For example,

$$3572·45 - 3558·61 = 13·84$$

Clearly it would be wrong to round off to the nearest integer during the calculation and say $3572 - 3559 = 13$. It is always safest to retain two or three more decimal places or significant figures during the calculation than will be required in the answer.

A common difficulty is that if a column of figures is rounded off and then added up, the result may not agree with the true total. Consider the following:

$$
\begin{array}{r}
19,247 \\
31,473 \\
67,285 \\
\hline
118,005
\end{array}
$$

If the three items are rounded off to the nearest thousand, the total becomes 117,000, but if the total is important, it should also be shown correct to the nearest thousand. The right way is not to 'force' the addition as in 'A' below, but to round off all items (including the total) independently as in 'B'.

| A. The wrong way *Thousands* | B. The right way *Thousands* |
|:---:|:---:|
| 19 | 19 |
| 32 | 31 |
| 67 | 67 |
| 118 | 118 |

## Spurious precision

In published statistics, such as trade returns, figures are often printed as recorded, not rounded off, even when they are subject to substantial revision. The reader is supposed to use his own discretion about rounding off or using them in further calculations. Care should be taken, however, to avoid giving the impression that the figures are more accurate than they really are. Thus it is legitimate to say 'the population of Slowe-on-the-Uptake as recorded in the last Census was 5267', or to say a few years after the Census that the population is 'about 5000'. It is not correct to say, years after the Census, 'the population is 5267'.

Foreign exchange is a common source of spurious precision. For example, it may be estimated that a fire in New York has caused damage to the value of a million dollars. In the good old days when the pound was worth \$2·40, this might have appeared as £416,667, whereas 'about £400,000' would be both safer and less misleading. The observant reader will detect many similar examples and should scrupulously avoid adding to them.

## Absolute and relative errors

The *absolute error* of an estimate is the actual difference between the estimate and the true value (which may be unknown), counted as positive or negative according to whether the estimate is greater or less than the true value. If $x$ represents the estimate and $X$ the true value, the error $e$ is given by:

$$e = x - X$$

The *relative error* is the absolute error divided by the true value, i.e. $e/X$. If $X$ is not known exactly, an estimate of it (not necessarily $x$ itself, but perhaps an average of several observations) can be used without increasing the error unduly. Relative errors are often expressed as percentage errors. For example, if 26,757 is rounded off to 27,000, the absolute error is −243 and the relative error is:

$$\frac{-243}{26,757} = -0·91 \text{ per cent}$$

## Combination of errors

The mathematical treatment of errors will be developed in later chapters, but the effect of errors in quantities subject to simple operations of arithmetic can be discussed briefly here. For simplicity, only two quantities will be considered, but the results can be generalized to any number.

Let $x_1$ and $x_2$ be the true values (assumed known) and $e_1$ and $e_2$ the errors, so that the observed values are $x_1 + e_1$ and $x_2 + e_2$.

Then

$$(x_1 + e_1) + (x_2 + e_2) = (x_1 + x_2) + (e_1 + e_2) \tag{6.1}$$

and

$$(x_1 + e_1) - (x_2 + e_2) = (x_1 - x_2) + (e_1 - e_2) \tag{6.2}$$

These equations show, if common sense does not, that in adding or subtracting quantities liable to error, the errors themselves must be added or subtracted in the same way to arrive at the error in the result.

The maximum possible errors in sums or differences can easily be found from this rule. Thus, if two amounts, £34,000 and £1400, are added together rounded off to the nearest £1000 and £100 respectively, the possible errors are £500 and £50, and the possible error in the result is £550. If the second item had been subtracted from the first, the possible error would still have been £550, not £450, since the two errors might be in opposite directions, e.g.

$$(34{,}000 - 500) - (1400 + 50) = 32{,}600 - 550$$

As the reader will see, the evaluation of possible errors in addition or subtraction only requires a little careful thought.

In multiplication and division it is the relative errors, not the absolute errors, that must be added or subtracted, since

$$(x_1 + e_1)(x_2 + e_2) = x_1 x_2 \left(1 + \frac{e_1}{x_1}\right)\left(1 + \frac{e_2}{x_2}\right)$$

$$\simeq x_1 x_2 \left(1 + \frac{e_1}{x_1} + \frac{e_2}{x_2}\right) \tag{6.3}$$

provided the product

$$\frac{e_1 e_2}{x_1 x_2}$$

can be neglected. (The sign $\simeq$ denotes approximate equality.) Similarly, it can be shown that

$$(x_1 + e_1) \div (x_2 + e_2) \simeq \frac{x_1}{x_2}\left(1 + \frac{e_1}{x_1} - \frac{e_2}{x_2}\right) \tag{6.4}$$

It follows from (6.3) that if a number is squared, any relative error in it is doubled. Conversely, when a square root is calculated, the relative error is halved.

It must be remembered that in all the above formulae $e_1$ and $e_2$ may be either positive or negative. In calculating possible, as distinct from actual, errors, the magnitudes of the possible errors must be added together as if they were all positive. Thus, if an operator performs the operation $a \times b \div c$ on a slide rule and reads each time to an accuracy of 0·4 per cent, the result will be accurate within 1·2 per cent. It is, of course, extremely unlikely that he would make the maximum error each time, but that is another matter.

## Exercises

6.1   Explain the terms

(*a*) Absolute error.
(*b*) Relative error.
(*c*) Percentage error.

The three dimensions of a rectangular block of stone are 1, 2 and 3 feet respectively, with a relative error of 0·01 in each case. What is the greatest possible percentage error in the volume stated as 6 cubic feet, and what is the maximum absolute error?                                     (*IoS*)

6.2   $S = P + Q$: $S$ is estimated at 156,000 correct to the nearest 1000, and $P$ at 49,500 correct to the nearest 100. Find the value of $S/Q$ with the possible limits of error.                                     (*IoS*)

6.3   You have to report some figures which are accurate only within certain limits. Give examples showing *three* ways of indicating this fact for one of the figures. Why do such margins of error arise in figures?

The following figures are accurate to the nearest unit only. Work out the limits of error.

$$\frac{\left(19 - \dfrac{17}{4}\right)3}{11}$$

(*ICMA*)

6.4   For three different areas, the number of households, and the average consumption of coal per household are given below. Make an estimate of the total consumption (*a*) for each separate dis-

| District | No of households | Tons/household |
|----------|------------------|----------------|
| A | 435,000 | 1·56 |
| B | 273,000 | 1·29 |
| C | 527,000 | 2·17 |

trict and (*b*) for the three districts combined, bearing in mind that the number of households is given to the nearest thousand and the average consumption to the nearest hundredth of a ton.                    (*IoS*)

6.5   Explain the method by which statistical figures are rounded and the reasons for doing so.

Round the commodity export values on the following table to the nearest £1000 and indicate the errors of approximation in the rounded figures.

| Commodity | Original values |
|-----------|-----------------|
| | £ |
| Spirits | 11,361,617 |
| Beer | 1,142,896 |
| Fruit juice and table waters | 223,490 |
| Cocoa preparations: | |
|     With sugar | 547,527 |
|     Without sugar | 264,771 |
| All other items | 283,561 |
| *Total* | 13,823,862 |

(*ACA*)

6.6   A motorist whose car has a broken distance recorder and who, therefore, measures his distances by map-reading, wishes to assess his petrol consumption on a Continental trip. His petrol gauge indicates his fuel consumption in litres with an error of $\pm 10$ per cent. He also uses approximate conversion factors of 1 gallon = 4·5 litres and 1 mile = 1·6 kilometres, instead of the more accurate 1 gallon = 4·55 litres and 1 mile = 1·61 kilometres. Estimate his probable petrol consumption in miles per gallon, giving limits of error, if he claims to be getting 31 miles per gallon after a trip of exactly 615 miles.           (*IoS*)

6.7   Explain the terms: (i) absolute error; (ii) relative error; (iii) biased error.

A manufacturer is planning to make a batch of new product. He estimates that: the material will cost £2000, but there may be a 10 per cent increase in the cost of materials; the wages will be £3000, but he may have to give his workers a 5 per cent pay increase.

He plans to produce 1000 items (to the nearest 10) and that the selling price per item will be £8 (to the nearest £1).

Calculate:

(i) Maximum cost of producing the items.
(ii) Maximum and minimum income from sales.
(iii) Maximum and minimum profit.
(iv) Maximum and minimum profit per item.                    (*IoS*)

# 7 Charts

The great advantage of charts is that the salient features stand out immediately. They add nothing to the information contained in the statistics they illustrate, but they bring out clearly comparisons and trends which the figures themselves might only reveal under close scrutiny. There are, however, certain limitations. The number of items depicted must be small, or confusion will result, and there must be enough movement to give sufficient contrasts. Also, charts require considerable skill in drawing and are expensive to reproduce, especially in more than one colour.

## Types of charts

Most readers will be familiar with the commoner types of chart, particularly with *graphs* in which continuous lines or curves depict relations between two variables (generally denoted by $x$ and $y$), represented on the graph by distances from two lines (the $x$- and $y$-axes) at right angles to each other. The student who needs to refresh his memory on the subject should refer to any elementary textbook on algebra.

Current usage does not distinguish clearly between the terms *diagram*, *figure* and *chart*. For convenience, the word *diagram* will be used to denote the simpler types of chart, generally involving lines, rectangles or circles. The words *chart* and *figure* can be used indifferently of both graphs and diagram.

Most of the charts occurring in statistics are either graphs or similar to graphs in representing relations between two variables. For example, there is the time-series chart, showing how one variable, such as output, sales, population or prices, varies with time. More will be said about this later in the present chapter. There is the frequency diagram, to be discussed in detail in Chapter 9, showing how frequency of occurrence varies with the value of some numerical characteristic. There is the scatter diagram, showing how one quantity depends partly on the value of another, e.g. a man's height and weight; this will be discussed in Chapter 14. There are also simpler diagrams, now to be described, generally designed for a wider public.

## Pie charts

The 'pie-chart' is a popular device for showing how an aggregate is divided into its principal components. Fig. 2 shows how the turnover of a company might be divided into cost of materials, wages and salaries, other costs, and profit.

**Fig. 2**   Analysis of turnover of Dunn, Wright and Co Ltd

In this diagram, the various items are proportional to the areas representing them in the circle, or to the angles of the various sectors. They can be converted into percentages by dividing the angles at the centre by 360 and multiplying by 100. If it is desired to represent absolute magnitudes, other methods are preferable, e.g. bar charts.

## Bar charts

In the simplest form of bar chart, several items are shown graphically by horizontal or vertical bars of uniform width with lengths proportional to the values they represent. Fig. 3 shows a simple example of this.

If preferred, the actual figures can be omitted, and the countries can

**Fig. 3** Density of population in various countries in 1968

be given in a key similar to that shown in Fig. 2, the bars being shaded or coloured correspondingly.

A rather more complicated chart, combined with a table, is given in Fig. 4(a), representing the coal output of the three main producing countries in different years. An alternative method of presenting the same data is shown in Fig. 4(b). Whereas Fig. 4(a) brings out the increase or decrease in each country's output, Fig. 4(b) focuses attention on the relative importance of the three countries at different times. It

| Country | Year | Million tonnes | Million tonnes |
|---------|------|----------------|----------------|
| UK | 1953 | 228 | |
| | 1963 | 199 | |
| | 1973 | 132 | |
| USA | 1953 | 440 | |
| | 1963 | 430 | |
| | 1973 | 530 | |
| USSR | 1953 | 218 | |
| | 1963 | 395 | |
| | 1973 | 461 | |

**Fig. 4(a)** Coal output of chief producing countries in 1953, 1963 and 1973

**Fig. 4(*b*)**  Coal output of chief producing countries in 1953, 1963 and 1973 (alternative form)

is irrelevant for this purpose whether the bars are horizontal or vertical, how the figures are shown (if at all), whether the bars are contiguous or separated, and whether the countries are distinguished by different shading or, better still if possible, by different colours. All these are largely matters of taste and money.

When it is desired to show an aggregate and its division into components, the bars are drawn proportional in length to the totals and divided in the ratios of their components, as in Fig. 5.

A loss appears in this diagram as a negative profit below the base line. This is included in the section representing 'other costs', but it is difficult to make this clear in the diagram. If all that is required is the division of the total into percentages this can be shown by a percentage bar chart as in Fig. 6.

**Fig. 5**  Analysis of the turnover of Sherlock's Salted Milk Ltd, 1975 and 1976

**Fig. 6**   Percentage analysis of UK imports by value, 1970 and 1975

The principle of the percentage bar chart is the same as that of the 'pie chart', and there is little to choose between them except that on the former the actual percentages are more easily shown and read off.

### Graphs of time series: line charts

The commonest type of chart in economic and commercial statistics is that of the time series, showing the progress of one or more quantities at successive times, usually at regular intervals. There are, in fact, although the distinction is often blurred, two kinds of time series. In one, the quantity studied is one which has a certain value at any given moment, e.g. population or stocks; in the other, it is a total or average relating to periods of time, e.g. annual or monthly production, average prices or weekly earnings. In practice, the distinction is not very important, since both kinds of series may be illustrated by the same kinds of charts.

Examples of time series have already been hinted at in Figs. 4(*b*) and 5, except that two years, or three at regular intervals, can hardly be said to constitute a series. If the data for these diagrams were continued for several years, either would provide a good example of a time series illustrated by a bar chart. It is simpler and usually more convenient, however, to use a line chart, in which points representing the data are joined by lines to aid the eye. Fig. 7 shows such a chart for production of cars in the United Kingdom from 1969 to 1975 with an earlier year (1956) for comparison. Notice that 1956, being so far removed in time from 1969, is joined to it by a broken line, whereas

**Fig. 7**  UK production of passenger cars, 1956 and 1969–75

**Fig. 8**  Imports into the UK, 1965 and 1973–6 (monthly averages or calendar months) (from Table 4)

**Table 4**  Imports into the UK, 1965 and 1973–1976
(monthly averages or calendar months)

| | £ million | | | £ million |
|---|---|---|---|---|
| 1965 | 479 | January | 1976 | 2181 |
| 1973 | 1321 | February | ,, | 2250 |
| 1974 | 1936 | March | ,, | 2181 |
| 1975 | 2014 | April | ,, | 2501 |
| | | May | ,, | 2577 |
| | | June | ,, | 2668 |
| | | July | ,, | 2704 |
| | | August | ,, | 2580 |
| | | September | ,, | 2815 |
| | | October | ,, | 2872 |

1969 to 1975 are equally spaced horizontally and joined by continuous lines.

A rather more complicated example is given in Table 4 and Fig. 8. Here it was desired to show the annual value of imports into the United Kingdom from 1973 to 1975 and monthly values thereafter, with an earlier year (1965) for comparison. In a case like this, where annual and monthly data appear together, it is necessary to give the

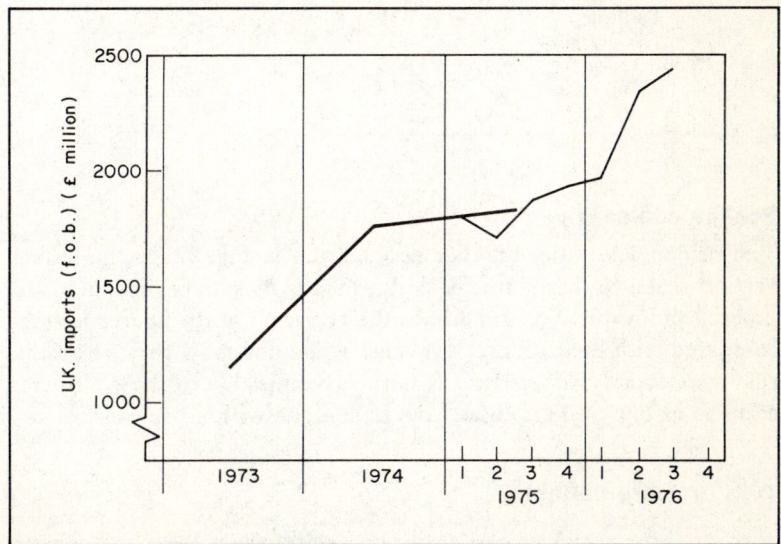

**Fig. 9**  Imports into the UK (fob), 1973–6 (monthly averages) (from Table 5)

annual data as monthly averages, and this should always be made clear.

Opinions will differ about the best way to present these data. The aim in Fig. 8 has been to suggest methods not shown earlier. The 1965 average appears as a broken horizontal line, and the change of scale necessitated by monthly data is marked by a vertical line and a broken line from the 1975 point to that for January 1976. If, however, it is intended eventually to show the average for 1976, the scale should be uniform throughout, even if the months are cramped together.

Fig. 9 shows how annual and quarterly data can be presented together; to make them comparable, they are all given as monthly averages. The lines joining quarterly points are lighter than those joining annual points, and the average for 1975 is shown midway between the second and third quarters.

**Table 5**   Imports into the UK, 1973–6 (fob)

(monthly averages)

|  |  | £ million |
|---|---|---|
| 1973 |  | 1175 |
| 1974 |  | 1760 |
| 1975: | 1st qr | 1794 |
|  | 2nd qr | 1720 |
|  | 3rd qr | 1874 |
|  | 4th qr | 1936 |
|  | Average | 1831 |
| 1976: | 1st qr | 1970 |
|  | 2nd qr | 2344 |
|  | 3rd qr | 2442 |

## Broken scales

The reader will notice another new feature in Fig. 9, viz. the broken vertical scale. Normally it is desirable for the scale to begin at zero and proceed uniformly, but sometimes the variation in the figures is small compared with their general level and would not show very well if this rule were strictly adhered to. It is then permissible to show a broken scale as in Fig. 9. This should never be done with a bar chart.

## Two or more variables

Two or three variables can easily be exhibited on the same chart if they are expressed in the same units. Thus exports could be shown in Fig. 8 by a second line distinguished from that for imports by being heavier or discontinuous. If coloured lines are used, four or five variables may be shown, e.g. sales of different commodities, but more than that would make the chart very confusing, especially if the lines kept crossing.

Variables in different units create further complications. If necessary, two different scales can be shown, one on the left and one on the right as in Fig. 10, the choice of relative scales being a matter

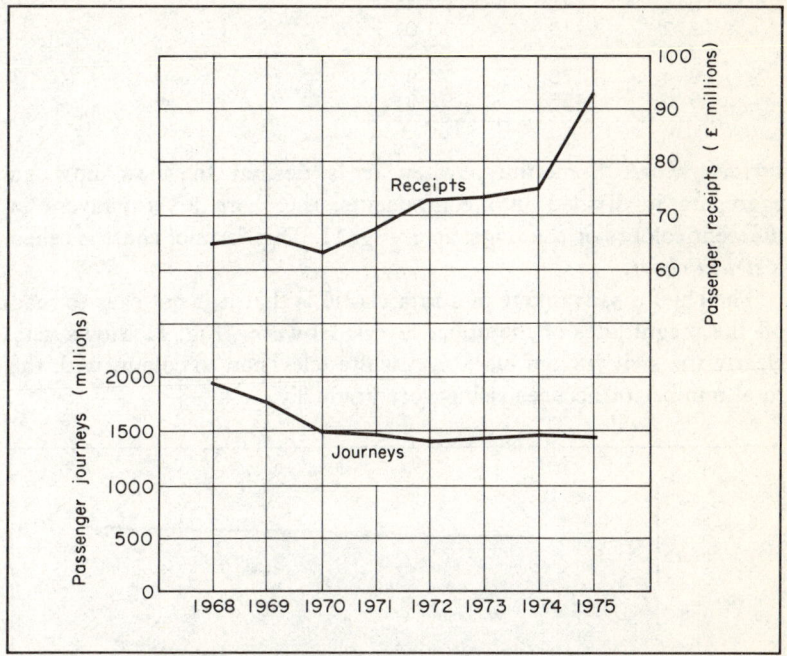

**Fig. 10** Passenger journeys and receipts on London Transport, 1968–75 (from Table 6)

for individual judgment. If colours can be used, each scale can be shown in the same colour as the corresponding variable. The figure will look better and be easier to read if the scales are chosen so that one is a simple multiple of the other. Alternatively, ratio scales may be used (see below).

## Surface and strata charts

When the space between the line or curve and the base line is filled in with some kind of colour or shading, the presentation is often called a

**Table 6**    Passenger journeys and receipts
on London Transport, 1968–75

| Year | Passenger journeys (millions) | Passenger receipts (£ millions) |
|------|------|------|
| 1968 | 1946 | 65·2 |
| 1969 | 1786 | 66·4 |
| 1970 | 1502 | 63·2 |
| 1971 | 1480 | 67·7 |
| 1972 | 1413 | 73·0 |
| 1973 | 1439 | 73·7 |
| 1974 | 1473 | 74·9 |
| 1975 | 1455 | 92·6 |

*surface chart*. Sometimes, when it is desired to show how an aggregate is divided into components, there are several layers in different colours or shadings, as in Fig. 11. This form of chart is called a *strata chart*.

The chief disadvantage of strata charts is that it is not easy to read off the magnitudes of the upper layers. However, Fig. 11 shows very clearly the switch from black and white television to colour, with the total number of licences rising very gradually.

**Fig. 11**    Television licences current in the UK, 1971–5

### Ratio or logarithmic charts

For some purposes it is better to use the scale obtained by plotting log *y* instead of *y*: the result is known as a Ratio Scale or Logarithmic Scale. Readers familiar with the theory of logarithms will realize that the effect of the change is to compress the grid in the upper part of the graph and expand it in the lower part, so that equal vertical intervals as measured on the graph now represent not equal absolute changes, but equal percentage changes. Thus a rise from 1 to 2, from 3 to 6, or from 500 to 1000 corresponds to the same vertical interval on the chart. It will be evident that the scale cannot start at zero, as a uniform scale normally does. If it starts at 1, the vertical distance of any point from the base line is proportional to the logarithm of the number it represents; hence the name. A slide rule is graduated in exactly the same way.

This method is extremely useful when it is desired to study relative rather than absolute movements, when a series contains both very small and very large values, or when the series is one that might be expected to increase roughly in geometric progression, like population. A series increasing or decreasing in this way, e.g. a sum of money accumulating at a constant rate of compound interest, is represented on a logarithmic chart by a straight line.

Fig. 12 shows the data of Table 6 and Fig. 10 plotted on a

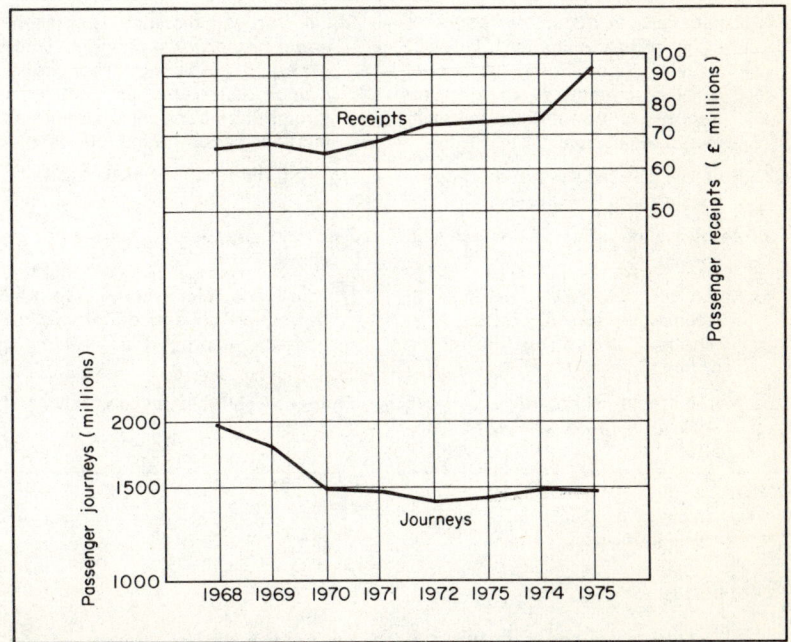

**Fig. 12** Passenger journeys and receipts on London Transport, 1968–75 (semi-logarithmic chart)

logarithmic chart (or strictly speaking, on a semi-logarithmic chart, for only the vertical scales are logarithmic, time being shown on a uniform horizontal scale). As with most ordinary line charts, the sole purpose of the lines is to aid the eye.

Special graph paper, known as semi-logarithmic paper, is available for these charts. It is scaled and ruled on the same principle as Fig. 12 and commonly gives two or three 'cycles' of 1 to 10, e.g. 'three-cycle' paper goes from 1 to 1000. If such paper is not to hand, a logarithmic chart can be drawn by plotting the logarithms of the values on ordinary graph paper.

The chief characteristics of natural scale and ratio scale graphs can be summarized as follows:

## Comparison of natural scale and ratio scale (logarithmic) graphs

| Natural scale | Ratio scale |
|---|---|
| 1. Equal vertical distances represent equal *absolute* changes. Lines of equal slope represent equal rates of change. A straight line denotes a continuous increase at simple interest. | Equal vertical distances represent equal *proportional* changes. Lines of equal slope represent equal *proportional* rates of change. A straight line denotes a continuous increase at compound interest. |
| 2. Suitable for analysing an aggregate into its constituents. | Not suitable. |
| 3. Zero and negative values may be shown. | Zero and negative values cannot be shown. |
| 4. Must be definitely located with reference to a base line, whether shown on the graph or not. | No base line. The whole curve may be moved up and down without affecting its properties. |
| 5. Not suitable when the y variable shows a great range of variation. | Eminently suitable in such cases. |

## Exercises

7.1 Illustrate the following data by means of bar charts *or* pie charts. Give the reasons for your choice of chart and state the relative advantages and disadvantages of the two methods of presentation.

Pens and pencils (manufacturers' sales) (£ thousand)

|  | *1956* | *1961* | *1966* |
|---|---|---|---|
| Fountain pens | 2973 | 2597 | 3236 |
| Ball point pens | 2361 | 3405 | 4535 |
| Felt tip pens | — | — | 677 |
| Fibre tip pens | — | — | 566 |
| Pencils | 1788 | 2013 | 1985 |

(Source: *Board of Trade*)                                      (*ICMA*)

7.2    As an accountant you have been asked by your Board of Directors to comment on the relative performance of the five firms making up the industry. The only available information is as follows:

|  | Quantity sales | | Turnover £ (000's) | |
|---|---|---|---|---|
|  | *1963* | *1964* | *1963* | *1964* |
| Firm A | 50,000 | 60,000 | 1600 | 1860 |
| ,,   B | 40,000 | 56,000 | 1240 | 1680 |
| ,,   C | 10,000 | 15,000 | 300 | 450 |
| ,,   D | 30,000 | 30,000 | 960 | 930 |
| ,,   E | 20,000 | 26,000 | 640 | 806 |

Assume all quantity sales are comparable.

Comment suitably to your Board and, in replying, any graphs or charts need not be accurate but must indicate salient features.

(*ACA*)

7.3    Draw charts to bring out the main features shown by the following figures of passenger movement. Say *briefly* what these main features are.

Destination of passengers leaving the UK

|  | Thousands | | | | |
|---|---|---|---|---|---|
|  | *1961* | *1962* | *1963* | *1964* | *1965* |
| *By sea* | | | | | |
| European and Mediterranean countries | 2486 | 2631 | 2915 | 3024 | 3295 |
| North America | 121 | 128 | 127 | 115 | 108 |
| Rest of world | 159 | 159 | 169 | 169 | 154 |
| *By air* | | | | | |
| Europe and Mediterranean countries | 2477 | 2716 | 3030 | 3400 | 3862 |
| North America | 398 | 459 | 528 | 670 | 782 |
| Rest of world | 164 | 183 | 210 | 260 | 322 |

(Source: *Annual Abstract of Statistics*)                    (*IoT*)

7.4   *Forecast car registrations over the period March 1966 to February 1967*

| Month | Actual registrations (thousands) | Predicted registrations (thousands) |
|---|---|---|
| March 1966 | 137·2 | 135·4 |
| April | 137·1 | 115·4 |
| May | 120·3 | 100·0 |
| June | 102·7 | 99·7 |
| July | 99·7 | 95·5 |
| August | 59·4 | 57·6 |
| September | 54·6 | 59·3 |
| October | 53·0 | 56·7 |
| November | 52·8 | 49·6 |
| December | 39·3 | 46·4 |
| January 1967 | 103·6 | 89·3 |
| February | 92·9 | 92·9 |

(Source: *The Statistician*, Vol. 17, No. 3, 1967)

Plot these two sets of figures on a graph and comment on the result.

(*IoS*)

7.5   'In 1912 there were 250,000 road vehicles in Britain. Twenty years later there were ten times as many. By 1952 the number had doubled again, to double yet again by 1962 when there were 10½ millions. There is now the prospect of 18 million vehicles by 1970, 25 millions by 1980 and more than 30 millions by the end of the century.' (*The Guardian*, 7 November 1966.)

Draw a graph to illustrate this statement and comment on the shape of the graph.

(*IoS*)

7.6   The table below shows the turnover from menswear by type of retail outlet in percentages for the summer season 1971 in Great Britain.

Draw bar-charts to illustrate the data.

| | Suits | Jackets | Trousers | Coats |
|---|---|---|---|---|
| Textile retailers | 76 | 73 | 69 | 72 |
| Chain stores | 6 | 10 | 11 | 10 |
| Mail Order Houses | 7 | 8 | 8 | 12 |
| Other outlets | 11 | 9 | 12 | 6 |

(Source: *Clothing Institute Journal*, July–August 1972)

7.7 A questionnaire has been designed to determine the film preferences of cinemagoers. Give a description of the statistical techniques you would use to illustrate the results diagrammatically. The questionnaire could include such topics as: age, sex and marital status of cinemagoers; preference for westerns, musicals, drama etc.; number of films seen in previous month. (N.B. You are *not* required to design the questionnaire.) (*IoS*)

7.8 Using a slide rule or a table of logarithms you are required to construct semi-log graphs and plot on them the following data:

In Great Britain

| Year | Total expenditure on highways | Total number of cars licensed | Total number of goods vehicles licensed | Total casualties in road accidents |
|------|------|------|------|------|
| | £ million | thousands | thousands | thousands |
| 1959 | 228·0 | 4,972 | 1,378 | 333 |
| 1960 | 238·0 | 5,532 | 1,448 | 348 |
| 1961 | 270·7 | 5,983 | 1,503 | 350 |
| 1962 | 301·2 | 6,560 | 1,522 | 342 |
| 1963 | 342·4 | 7,380 | 1,582 | 356 |
| 1964 | 405·8 | 8,252 | 1,633 | 385 |
| 1965 | 421·2 | 8,922 | 1,661 | 397 |
| 1966 | 457·4 | 9,522 | 1,639 | 392 |
| 1967 | 528·2 | 10,312 | 1,692 | 370 |
| 1968 | 580·9 | 10,825 | 1,640 | 349 |

(Source: *Annual Abstract of Statistics*)

Interpret the graphs you have drawn and give the advantages, if any, of this type of presentation. (*ICMA*)

# 8 Analysis of time series

This chapter deals with elementary methods of analysing time series. A time series is constructed by plotting successive values of some variable ($x$) at successive intervals of time ($t$). It is implied that this series is long enough to enable vital statistical inferences to be drawn. The most interesting series are those of 'economic indicators', i.e. those which measure aspects of current economic activity as represented by statistics of prices, production, transport, bank clearings and so on.

In general, $x$ is a composite quantity: it represents the result of a number of influences which cannot be isolated for purposes of study and measurement. Nevertheless, statistical methods provide a means of estimating the contributions of the more important factors, and this is the problem to be considered.

According to current practice, a time series may be regarded as involving as many as five components, namely:

1. The *Trend*, or the course the series would take over a long period in the absence of disturbing factors.

2. *Cyclical fluctuations*, or wave-like disturbances corresponding, for example, with the movements of the Trade Cycle or the stock market.

3. *Seasonal variations* associated with the harvests, the weather, Christmas and other festivals, and possibly the varying length of the month, although this is of minor importance.

4. *Catastrophic movements* caused by unusual or unexpected events, e.g. the Suez crisis of 1956 or the severe winter of 1963.

5. *Residual variations*, which include all movements not already covered in headings 1 to 4.

For elementary purposes it will be sufficient to consider three components, namely: the trend (1), seasonal variations (3), and deviations from the trend—the result of pooling (2), (4) and (5)—and the rest of this chapter has been written on these lines.

Undoubtedly there are seasonal effects in many time series, although it is generally difficult to estimate them numerically. It is also, as a rule, possible to detect a general trend over a period, although the short-term trend may be quite different from the long-term trend. There is no reason why the above analysis should not be used to explain current events, but to use it for forecasting is hazardous because much depends on politics or unforeseen circumstances. If, however, there is a well-defined trend over a long period, it may be reasonable to extrapolate into the not-too-distant future.

## Finding the trend

Various methods have been devised for breaking up series into the elements mentioned above. It is generally most important to find the trend, or perhaps it would be better to say *a* trend, as there is seldom an exact and unique solution to this problem.

It is always helpful to plot the series on a graph, either on a natural scale or on a semi-logarithmic scale. The latter will be better if there is a tendency for the series to increase or decrease in geometric progression, as the trend will be more like a straight line on a logarithmic chart than on uniform graph paper. Assuming that, in the long run, deviations from the trend will be as often positive as negative, and vice versa, the trend can be estimated by any of the following methods:

1. Drawing a smooth freehand curve so that about the same number of points lie on either side of it. This method demands greater skill and judgment than are possessed by the average beginner, and even trained operators may differ considerably in their results.

2. Fitting a straight line or a mathematical curve to the points by methods to be described in Chapter 14.

3. Calculating a suitable *Moving Average* (see below).

## Moving average

This device consists in substituting for the original series a smoother series, each term of which is the average of a number (always the same number) of terms of the original series. Thus, if the original series is:

81, 92, 100, 106, 116, 127 etc.

and a 5-point moving average is required, the first term will be:

$$\tfrac{1}{5}(81 + 92 + 100 + 106 + 116) = 99$$

and the second:

$$\tfrac{1}{5}(92 + 100 + 106 + 116 + 127) = 108\cdot2,$$

and so on. At each step, one term of the original series is dropped and another is introduced. In fact, the second average could be obtained simply by adding $\tfrac{1}{5}(127 - 81) = 9\cdot2$ to the first. Table 7 shows how a 5-point moving average is calculated for a series of weekly data, and Fig. 13 shows the result plotted on a graph.

**Table 7**   Percentage time lost through sickness in a factory over a period of 21 weeks

| Week | Percentage time lost | Sum of 5 | Moving average |
|------|------|------|------|
| 1  | 2·5 |      |     |
| 2  | 3·3 |      |     |
| 3  | 4·0 |      | 4·1 |
| 4  | 4·7 |      | 4·6 |
| 5  | 6·0 | 20·5 | 4·7 |
| 6  | 5·0 | 23·0 | 4·6 |
| 7  | 3·6 | 23·3 | 5·2 |
| 8  | 3·7 | 23·0 | 5·6 |
| 9  | 7·8 | 26·1 | 5·5 |
| 10 | 7·7 | 27·8 | 5·4 |
| 11 | 4·7 | 27·5 | 5·3 |
| 12 | 3·0 | 26·9 | 4·1 |
| 13 | 3·2 | 26·4 | 3·3 |
| 14 | 2·1 | 20·7 | 2·5 |
| 15 | 3·3 | 16·3 | 2·0 |
| 16 | 1·1 | 12·7 | 1·5 |
| 17 | 0·4 | 10·1 | 1·3 |
| 18 | 0·7 | 7·6  | 1·1 |
| 19 | 0·8 | 6·3  | 1·3 |
| 20 | 2·4 | 5·4  |     |
| 21 | 2·4 | 6·7  |     |

Each sum of five items after the first is obtained by adding to the previous sum of five the difference between the new term and the one dropped. Thus, for weeks 11–15, the sum is:

$$20\cdot7 + (3\cdot3 - 7\cdot7) = 16\cdot3$$

The final entry in the third column should be checked by direct addition and, if the series is a long one, a few intermediate checks should be provided.

Notice that each average is plotted in the centre of the period it covers. This is clearly right, for in the extreme case of a series in perfect arithmetic progression, the moving average 'centred' in this way would coincide with the original series, and both would be depicted by the same straight line. It means, however, that the moving average will miss nearly half the period (two years in the case of a 5-year average) at each end.

**Fig. 13**   Percentage time lost through sickness in a factory (from Table 7)

As will be seen in Fig. 13, the moving average still shows kinks, although it is much smoother than the original series. It can, if desired, be further smoothed by a freehand curve or by a further averaging process.

Where there are cyclical fluctuations, the number of terms in the moving average will generally depend on the period of the cycle to be eliminated. Thus a 7-year cycle may be eliminated by a 7-year (or perhaps a 14- or 21-year) average, and seasonal movements by a moving average based on twelve months or four quarters. In the latter case, in fact whenever the moving average has an even number of terms, it can still be plotted in the middle of the period although it will now lie between the two middle items, not in the same vertical as a single middle item. There is no great objection to this, but for the purpose of trying to measure seasonal variations it is desirable for the moving average to relate to the same time or period as one of the original items. This can be arranged by averaging a step further. Suppose, for example, it is desired to plot the trend, and seasonal deviations from the trend, of a series of quarterly data. The average of the first four quarters will be midway (in time) between the second and third, and the average of the second to fifth quarters midway between the third and fourth. Clearly, then, the average of those two averages will give the trend for the third quarter. It is most simply obtained by first calculating the sums of four quarters, or *moving annual totals*, then adding those in pairs and dividing by 8, as in Table 8. For monthly data the procedure would be similar, but each MAT would be the sum of twelve items and the divisor would be 24 instead of 8.

**Table 8**   Sales of a certain fertilizer by Gromore and Sons Ltd

(Thousand tons)

| Year and quarter | | Sales | Moving annual total | Sum of two MATs | Moving average | Deviation from trend |
|---|---|---|---|---|---|---|
| 1972: | 1 | 60 | | | | |
| | 2 | 65 | | | | |
| | 3 | 20 | | | 47 | −27 |
| | 4 | 44 | 189 | | 47 | − 3 |
| 1973: | 1 | 62 | 191 | 380 | 47 | +15 |
| | 2 | 58 | 184 | 375 | 49 | + 9 |
| | 3 | 28 | 192 | 376 | 52 | −24 |
| | 4 | 50 | 198 | 390 | 53 | − 3 |
| 1974: | 1 | 85 | 221 | 419 | 52 | +33 |
| | 2 | 42 | 205 | 426 | 52 | −10 |
| | 3 | 33 | 210 | 415 | 55 | −22 |
| | 4 | 44 | 204 | 414 | 63 | −19 |
| 1975: | 1 | 118 | 237 | 441 | 65 | +53 |
| | 2 | 71 | 266 | 503 | 65 | + 6 |
| | 3 | 20 | 253 | 519 | 66 | −46 |
| | 4 | 58 | 267 | 520 | 66 | − 8 |
| 1976: | 1 | 110 | 259 | 526 | 68 | +42 |
| | 2 | 83 | 271 | 530 | 68 | +15 |
| | 3 | 22 | 273 | 544 | | |
| | 4 | 55 | 270 | 543 | | |

## Seasonal variations

When a series is strongly seasonal, like sales of ice-cream or the number of houses built each month, the fluctuations are apt to be confusing, and various methods have been devised for removing them so that one month may legitimately be compared with another. The best and most straightforward way of doing this is shown in Tables 8, 9 and 10, with quarterly data for simplicity. Monthly data can be treated on the same lines. It must be stressed, however, that this procedure is only legitimate when seasonal movements are very marked and reasonably consistent from year to year.

**Table 9** Fertilizer sales—seasonal movements

(Derived from Table 8)

| Year | Quarter | | | | Total |
|------|---------|---|---|---|-------|
| | *1* | *2* | *3* | *4* | |
| 1972 | | | − 27 | − 3 | |
| 1973 | + 15 | + 9 | − 24 | − 3 | |
| 1974 | + 33 | −10 | − 22 | −19 | |
| 1975 | + 53 | + 6 | − 46 | − 8 | |
| 1976 | + 42 | +15 | | | |
| Total | +143 | +20 | −119 | −33 | +11 |
| Adjusted total | +140 | +17 | −121 | −36 | 0 |
| Average | + 35 | + 4 | − 30 | − 9 | 0 |

**Table 10** Fertilizer sales corrected for seasonal movements

(Thousand tons)

| Year and quarter | Actual sales | Average seasonal movement | Sales (adjusted) |
|------------------|--------------|---------------------------|------------------|
| 1972: 1 | 60 | +35 | 25 |
| 2 | 65 | + 4 | 61 |
| 3 | 20 | −30 | 50 |
| 4 | 44 | − 9 | 53 |
| 1973: 1 | 62 | +35 | 27 |
| 2 | 58 | + 4 | 54 |
| 3 | 28 | −30 | 58 |
| 4 | 50 | − 9 | 59 |
| 1974: 1 | 85 | +35 | 50 |
| 2 | 42 | + 4 | 38 |
| 3 | 33 | −30 | 63 |
| 4 | 44 | − 9 | 53 |
| 1975: 1 | 118 | +35 | 83 |
| 2 | 71 | + 4 | 67 |
| 3 | 20 | −30 | 50 |
| 4 | 58 | − 9 | 67 |
| 1976: 1 | 110 | +35 | 75 |
| 2 | 83 | + 4 | 79 |
| 3 | 22 | −30 | 52 |
| 4 | 55 | − 9 | 64 |

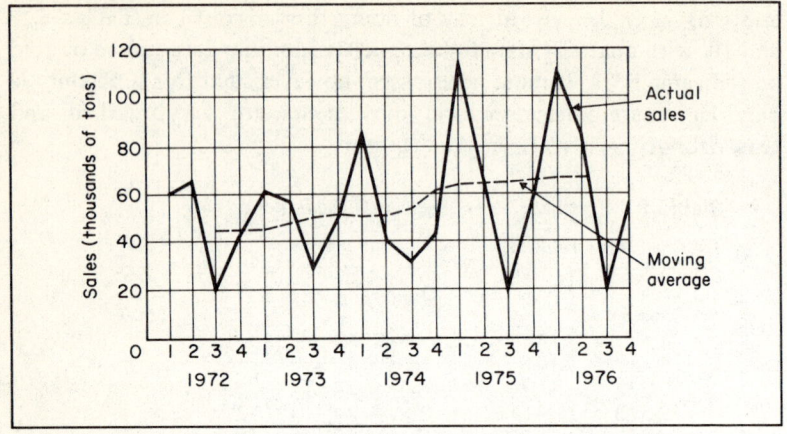

**Fig. 14**    Fertilizer sales by Gromore and Sons Ltd (from Table 8)

The moving average is first calculated as already described, and subtracted from the corresponding actual value to give the deviation from trend, e.g. $20 - 47 = -27$. These deviations are rearranged as in Table 9 and totalled for each month or quarter of the year. If the totals do not add up to zero, they should be adjusted as shown before the averages for each quarter are calculated. These averages give the best available estimates of the seasonal effects. Finally, these estimated seasonal fluctuations are subtracted from the original items as in Table 10 to obtain 'corrected' sales, i.e. what the sales would have been without the seasonal effect.

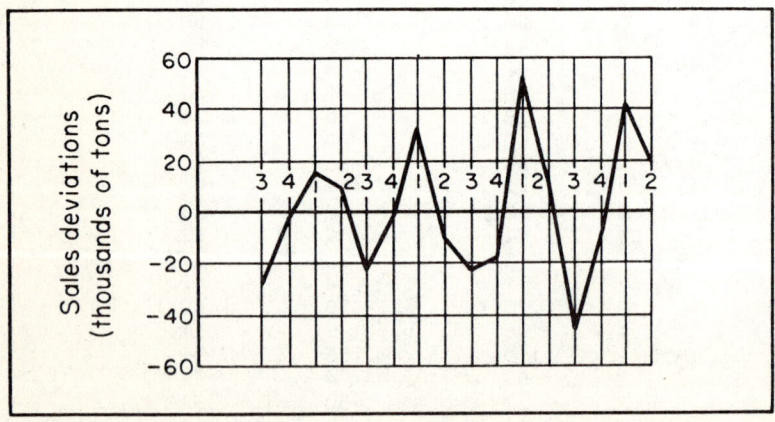

**Fig. 15**    Deviations from trend (from Table 8)

**Fig. 16** Fertilizer sales, corrected for seasonal movements (from Table 10)

The results are shown graphically in Figs. 14, 15 and 16. As generally happens, the deviations from trend are partly seasonal and partly what, for want of a better term, is described as 'random'. The adjusted series is smoother than the original but not as smooth as the trend line because, although the seasonal effect has been removed (as far as possible), the residual or random fluctuation is left in. The movements do not show the same regularity as in the original series.

Table 11 shows the final analysis of the original data into trend or

**Table 11** Analysis of fertilizer sales

*(Thousand tons)*                                                                          *(Data of Table 8)*

| Year and quarter | | Trend | Seasonal movement | Residual | Actual sales |
|---|---|---|---|---|---|
| 1972: | 3 | 47 | −30 | + 3 | 20 |
| | 4 | 47 | − 9 | + 6 | 44 |
| 1973: | 1 | 47 | +35 | −20 | 62 |
| | 2 | 49 | + 4 | + 5 | 58 |
| | 3 | 52 | −30 | + 6 | 28 |
| | 4 | 53 | − 9 | + 6 | 50 |
| 1974: | 1 | 52 | + 5 | − 2 | 85 |
| | 2 | 52 | + 4 | − 4 | 42 |
| | 3 | 55 | −30 | + 8 | 33 |
| | 4 | 63 | − 9 | −10 | 44 |
| 1975: | 1 | 65 | +35 | +18 | 118 |
| | 2 | 65 | + 4 | + 2 | 71 |
| | 3 | 66 | −30 | −16 | 20 |
| | 4 | 66 | − 9 | + 1 | 58 |
| 1976: | 1 | 68 | + 5 | + 7 | 110 |
| | 2 | 68 | + 4 | + 1 | 83 |

moving average, seasonal movement, and residual. Since the moving average cannot be calculated for the two quarters at each end of the original period, these quarters are omitted from the analysis.

The method described above allows automatically for the varying length of the month and for most public holidays. A further adjustment may be necessary, however, if Easter occurs in April in one year and in March in another.

### An alternative model

So far we have assumed that, if there were no random or residual variation, the seasonal differences from trend would be constant. This is an example of what, in the modern idiom (some would say jargon), is called a *statistical model*. A model is a hypothesis, often expressed in the form of equations or other mathematical terms, which we think may adequately describe the underlying processes. Usually it is over-simplified, but it often fits the facts well enough to enable us to use it for forecasting and similar purposes.

**Table 12**  Sales by Gromore and Sons Ltd (alternative model)

| Year and quarter | Sales (1000 tons) | Moving average | Sales as percentage of moving average |
|---|---|---|---|
| 1972: 1 | 60 | | |
| 2 | 65 | | |
| 3 | 20 | 47 | 43 |
| 4 | 44 | 47 | 94 |
| 1973: 1 | 62 | 47 | 132 |
| 2 | 58 | 49 | 118 |
| 3 | 28 | 52 | 54 |
| 4 | 50 | 53 | 94 |
| 1974: 1 | 85 | 52 | 163 |
| 2 | 42 | 52 | 81 |
| 3 | 33 | 55 | 60 |
| 4 | 44 | 63 | 70 |
| 1975: 1 | 118 | 65 | 182 |
| 2 | 71 | 65 | 109 |
| 3 | 20 | 66 | 30 |
| 4 | 58 | 66 | 88 |
| 1976: 1 | 110 | 68 | 162 |
| 2 | 83 | 68 | 122 |
| 3 | 22 | | |
| 4 | 55 | | |

This particular model, however, may not be the best for a time series. If the general level of sales doubles over a period, it is likely that both seasonal and residual variations will, on the average, double also. In other words, if there is a definite trend upwards or downwards, it is unlikely that seasonal differences from trend will be constant; it is more reasonable to suppose that *seasonal factors* will be constant, i.e. that in the absence of residual variation, the ratio of any item to the corresponding trend value will be constant for any given month or quarter of the year.

**Table 13**  Calculation of seasonal factors

| Year | Quarter | | | | Total |
| | 1 | 2 | 3 | 4 | |
|---|---|---|---|---|---|
| 1972 | | | 43 | 94 | |
| 1973 | 132 | 118 | 54 | 94 | |
| 1974 | 163 | 81 | 60 | 70 | |
| 1975 | 182 | 109 | 30 | 88 | |
| 1976 | 162 | 122 | | | |
| Total | 639 | 430 | 187 | 346 | 1602 |
| Average | 160 | 107 | 47 | 86 | 400 |

Tables 12, 13 and 14 show how seasonal factors and seasonally corrected figures are obtained with this second model. First, each quarter's sales are expressed as a percentage of the moving average, e.g. for the third quarter of 1974:

$$\frac{33}{55} \times 100 = 60$$

Next these percentages are averaged in Table 13, just as the differences were in Table 9. If necessary, the totals should be adjusted so that the average seasonal factors (expressed as percentages) add up to 400 exactly; in this example it happened to be unnecessary, allowing for some rounding off. Finally, the actual sales are divided by the appropriate factors to obtain seasonally corrected figures.

Thus, for the third quarter of 1974:

$$\frac{33}{0 \cdot 47} = \frac{33 \times 100}{47} = 70$$

These seasonally corrected figures are shown in Fig. 17, which should be compared with Fig. 16. The residual variation is still large, but

**Table 14**   Sales by Gromore and Sons Ltd, seasonally corrected

| Year and quarter | | Actual sales (1000 tons) | Seasonal factor | Adjusted sales (1000 tons) |
|---|---|---|---|---|
| 1972: | 1 | 60 | 160 | 38 |
| | 2 | 65 | 107 | 61 |
| | 3 | 20 | 47 | 43 |
| | 4 | 44 | 86 | 51 |
| 1973: | 1 | 62 | 160 | 39 |
| | 2 | 58 | 107 | 54 |
| | 3 | 28 | 47 | 60 |
| | 4 | 50 | 86 | 58 |
| 1974: | 1 | 85 | 160 | 53 |
| | 2 | 42 | 107 | 39 |
| | 3 | 33 | 47 | 70 |
| | 4 | 44 | 86 | 51 |
| 1975: | 1 | 118 | 160 | 74 |
| | 2 | 71 | 107 | 66 |
| | 3 | 20 | 47 | 43 |
| | 4 | 58 | 86 | 67 |
| 1976: | 1 | 110 | 160 | 69 |
| | 2 | 83 | 107 | 78 |
| | 3 | 22 | 47 | 47 |
| | 4 | 55 | 86 | 64 |

rather less than in Fig. 16, which suggests that the second model is better than the first, at least in this example. It may not always be; each case must be judged on its own merits. In practice, the second model is usually preferred, but the first model is nearly always used in examination questions for the same reason as the data are generally

**Fig. 17**   Fertilizer sales, corrected for seasonal movements (alternative model) (from Table 14)

quarterly, not monthly: to keep the amount of calculation in one question within reasonable limits.

## Looking ahead

Forecasting is always hazardous—indeed, there are some who dismiss it contemptuously as 'mere crystal-gazing'. Admittedly it often goes wrong. At the time of writing, thousands of would-be teachers are facing unemployment largely because of over-optimistic forecasts of requirements made a few years ago. Weather forecasts are by no means infallible, and as for investment advisers and stockbrokers—!

However, if we are going to plan for the future at all, we have to take a view and risk making mistakes. Projecting forwards from past data is only one of several ways, but a useful one, provided that present trends can reasonably be expected to persist in the near future. Suppose Gromore and Sons want to forecast sales for each quarter of 1977, using our second model. They would first extend or *extrapolate* the trend from the last two or three years, probably fitting a straight line to the moving average. They would then multiply by the seasonal factors to obtain estimates of actual sales, as follows:

| Quarter | Trend (1000 tons) | Seasonal factor | Forecast sales (1000 tons) |
|---------|-------------------|-----------------|----------------------------|
| 1977: 1 | 71 | 160 | 114 |
| 2 | 72 | 107 | 77 |
| 3 | 73 | 47 | 34 |
| 4 | 74 | 86 | 64 |

These forecasts should be compared, if possible, with estimates obtained by other methods, or modified in the light of any special information regarding, say, Government policy or customers' stated intentions. Naturally, the further one looks into the future, the more uncertain these forecasts become.

## Z-charts

The Z-chart is so called because its three component curves roughly form the letter Z, or rather a succession of Zs. It can be adapted to any period, but each Z normally takes one year, and it will be assumed in what follows that this is the period covered by the Z.

The three components are:

1. The curve of the original data, generally but not always monthly.

2. The cumulative curve of the original data, e.g. the point for May 1975 will represent the total sales of, say, the five months January to May 1975.

3. The moving annual total, e.g. the point for May 1976 now represents the twelve months June 1975 to May 1976. The MAT is often extremely useful, quite apart from the Z-chart, as seasonal variations are automatically cancelled out.

**Table 15**   Sales of 'Boxo', 1975–7

| Year and month | | Monthly sales (£000) | Cumulative sales for year (£000) | Moving annual total sales (£000) |
|---|---|---|---|---|
| 1975: | January | 45 | 45 | |
| | February | 46 | 91 | |
| | March | 53 | 144 | |
| | April | 31 | 175 | |
| | May | 48 | 223 | |
| | June | 58 | 281 | |
| | July | 49 | 330 | |
| | August | 22 | 352 | |
| | September | 82 | 434 | |
| | October | 70 | 504 | |
| | November | 65 | 569 | |
| | December | 79 | 648 | 648 |
| 1976: | January | 62 | 62 | 665 |
| | February | 59 | 121 | 678 |
| | March | 76 | 197 | 701 |
| | April | 44 | 241 | 714 |
| | May | 54 | 295 | 720 |
| | June | 76 | 371 | 738 |
| | July | 54 | 425 | 743 |
| | August | 31 | 456 | 752 |
| | September | 68 | 524 | 738 |
| | October | 76 | 600 | 744 |
| | November | 83 | 683 | 762 |
| | December | 104 | 787 | 787 |
| 1977: | January | 97 | 97 | 822 |
| | February | 92 | 189 | 855 |
| | March | 105 | 294 | 884 |

Clearly if the same scale is used for all three curves the cumulative curve will coincide with (1) at the beginning of the year and with (3) at the end of the year, and the Z will be one continuous curve, as in Fig. 18. Sometimes, however, the monthly data are plotted on a larger scale than the other two (say five times as large) to bring them out more clearly. This is certainly necessary with weekly or daily data.

**Fig. 18**  Z-Chart representing sales of 'Boxo' (from Table 15)

## Exercises

8.1  Compute the trend from the following data, stating how you determine the period of cyclical fluctuation.

| Year | Sales per head (lb) | Year | Sales per head (lb) |
|------|------|------|------|
| 1 | 11·5 | 12 | 14·0 |
| 2 | 11·2 | 13 | 12·8 |
| 3 | 11·9 | 14 | 13·0 |
| 4 | 13·2 | 15 | 13·7 |
| 5 | 13·9 | 16 | 15·1 |
| 6 | 13·4 | 17 | 16·2 |
| 7 | 11·8 | 18 | 15·9 |
| 8 | 11·7 | 19 | 14·6 |
| 9 | 12·4 | 20 | 14·5 |
| 10 | 13·6 | 21 | 15·3 |
| 11 | 14·9 | 22 | 16·9 |

(*IoS*)

8.2   Describe concisely the steps involved in calculating, by the method of moving averages, a seasonally-adjusted series. Of what use is such a series in a transport undertaking? By analysing his receipts in recent years, a transport operator estimates that on average his takings each January are 20 per cent below trend, those in February are 10 per cent below, and those in March are 5 per cent below. In January, February and March 1963, his actual takings were £1000, £1100 and £1150 respectively. After allowing for seasonal influences, would you conclude that his receipts were rising or falling during this period?

*(IoT)*

8.3   The table below shows the sales of a certain article in each quarter over a period of 5 years. Using the method of moving averages, produce figures which show:

(*a*) The seasonal variation.
(*b*) The trend after eliminating seasonal variation.

| Year | 1 | 2 | 3 | 4 | 5 |
|------|-----|-----|-----|-----|-----|
| Spring | 58 | 79 | 97 | 113 | 136 |
| Summer | 29 | 48 | 66 | 91 | 105 |
| Autumn | 49 | 68 | 85 | 100 | 123 |
| Winter | 89 | 107 | 134 | 148 | 170 |

*(IoS)*

8.4   Use the method of moving averages to fit a trend line to the following quarterly figures of UK passenger movements by sea and air. Calculate seasonal adjustments. Draw a graph of the original data and the trend, and hence forecast UK passenger movements by sea and air for the 1st and 2nd quarters of 1968.

| Year | 1st quarter | 2nd quarter | 3rd quarter | 4th quarter |
|------|-----|-----|-----|-----|
| 1964 | 1023 | 2268 | 4383 | 1249 |
| 1965 | 1075 | 2615 | 4843 | 1367 |
| 1966 | 1201 | 2807 | 5363 | 1541 |
| 1967 | 1411 | 2938 | 5613 | 1708 |

*(IoT)*

8.5   Obtain an estimate of the normal seasonal movement in the series of unemployment figures shown below. If the unadjusted figure for the first quarter of 1976 is 870, obtain a seasonally adjusted figure, and state whether you believe that unemployment is on the increase.

No. of wholly unemployed (thousands) in Carpathia

|           | 1972 | 1973 | 1974 | 1975 |
|-----------|------|------|------|------|
| Quarter 1 | 620  | 663  | 704  | 748  |
| 2         | 630  | 661  | 690  | 731  |
| 3         | 610  | 631  | 644  | 720  |
| 4         | 674  | 712  | 799  | 868  |

*(IoS)*

8.6

Imports of Wine into Great Britain

| Quarters | *Million gallons* | | | |
|----------|------|------|------|------|
|          | *1*  | *2*  | *3*  | *4*  |
| 1970     |      |      |      | 13·6 |
| 1971     | 7·1  | 9·5  | 10·8 | 14·8 |
| 1972     | 8·8  | 11·9 | 12·3 | 17·8 |
| 1973     | 10·8 | 13·8 | 16·4 |      |

By means of a moving average, find the trend and the seasonal adjustments.

Forecast the wine imports for the fourth quarter of 1973.    *(IoS)*

8.7 (*a*)   Draw a Z-chart using the information given below.

(*b*) What information can a manager obtain from this chart which will help him improve his effectiveness?

Data, in units:

| Month | Year 1 for month | | Year 2 for month | | Year 2 Moving annual totals | |
|-------|------------|-------|------------|-------|------------|-------|
|       | *Production* | *Sales* | *Production* | *Sales* | *Production* | *Sales* |
| 1     | 239        | 346   | 244        | 336   | 2796       | 2834  |
| 2     | 248        | 332   | 257        | 299   | 2805       | 2801  |
| 3     | 213        | 185   | 230        | 163   | 2822       | 2779  |
| 4     | 236        | 169   | 242        | 195   | 2828       | 2805  |
| 5     | 239        | 168   | 215        | 155   | 2804       | 2792  |
| 6     | 229        | 177   | 176        | 101   | 2751       | 2716  |
| 7     | 206        | 195   | 203        | 113   | 2748       | 2634  |
| 8     | 250        | 189   | 200        | 201   | 2698       | 2646  |
| 9     | 216        | 177   | 198        | 178   | 2680       | 2647  |
| 10    | 251        | 318   | 264        | 345   | 2693       | 2674  |
| 11    | 235        | 335   | 215        | 343   | 2673       | 2682  |
| 12    | 229        | 253   | 216        | 175   | 2660       | 2604  |
| Totals | 2791      | 2844  | 2660       | 2604  |            |       |

*(ICMA)*

# 9 Frequency distributions

Frequency distribution is one of the fundamental ideas of statistics. It has already been noted that before a large number of observations can be analysed, or even made readily intelligible, they must be sorted into a convenient number of groups or classes. In the problems now to be discussed, the data are sorted according to the numerical value of some variable. Thus, a number of people might be sorted according to their height, weight, age, or any other characteristic capable of being measured. The number of items in any group is called the *frequency* of that group, and the resulting distribution is called (not unnaturally) the *frequency distribution*.

Table 16 shows a simple example of a frequency distribution, one of 500 families classified according to the number of children in a family.

## Continuous and discrete variables

A variable that can take only discrete values, as in the above example, is said to be *discrete* or *discontinuous*. Obviously a man cannot have 2·63 children or pay 9·748p for a bus fare. Other examples of discrete variables are the number of houses in a street or rooms in a house, the number of apples bought in 1 lb, and the price of an ice-cream.

A *continuous* variable, on the other hand, can take any value whatever within a certain range—or, more probably, an uncertain range. Thus, a temperature need not be an exact number of degrees, but may theoretically be measured to several decimal places, and any object whose temperature is rising will, during the process, take every possible temperature between the initial and final values. Similar examples are the speed of a vehicle and the height of a growing plant. In short, a continuous variable can take an infinite number of values.

The distinction between continuous and discrete variables, however, tends to become blurred. In the first place, it is often

**Table 16**  Frequency distribution
of 500 families by number of
children

| No of children in family | No of families |
|:---:|:---:|
| 0 | 179 |
| 1 | 205 |
| 2 | 78 |
| 3 | 26 |
| 4 | 7 |
| 5 | 3 |
| 6 | 1 |
| 7 | – |
| 8 | 1 |
| *Total* | 500 |

desirable to treat discrete variables with many possible values as 'almost continuous'. For instance, suppose a large firm records the earnings of each worker in a given week; the variable may be any whole number of pence up to, say, 8000. Strictly speaking, it is discrete, but the gap between successive values is so small and the number of possible values so large that the variable may be regarded for practical purposes as continuous. On the other hand, it is often necessary to treat continuous variables as discrete by rounding off values to the nearest inch, pound etc.; this can generally be done without appreciable loss of accuracy. As will be shown in this and subsequent chapters, it greatly facilitates work and simplifies the arithmetic.

## Grouping the data

In order to obtain a manageable frequency distribution when the variable is continuous, or exhibits a large number of discrete values, the range of possible values must be divided into a suitable number of sub-ranges and the data sorted accordingly. A group of men might be classified according to height and sorted into the following *groups*, or *classes*: under 5 ft; 5 ft and under 5 ft 2 in.; and so on (see Table 17).

The number of items in a particular class is called the *class-frequency*, and the range of the variable in that class the *class-interval*. In Table 17 the class-frequencies are 2, 5, 14 etc., and the class-interval is 2 inches throughout, except at the extremes, where there is no definite class-interval.

**Table 17**   Frequency distribution of 1000 men by height

| Height (inches) | No of men |
|---|---|
| Under 58 | 2 |
| 58 and under 60 | 5 |
| 60 and under 62 | 14 |
| 62 and under 64 | 60 |
| 64 and under 66 | 187 |
| 66 and under 68 | 304 |
| 68 and under 70 | 263 |
| 70 and under 72 | 121 |
| 72 and under 74 | 36 |
| 74 and under 76 | 7 |
| 76 and over | 1 |
| *Total* | 1000 |

It is usual to have a uniform class-interval throughout, as this gives a much clearer picture of the distribution than class-intervals of varying size. The reader can readily imagine what a confused impression Table 17 would have given if the men had been grouped as follows: under 5 ft; 5 ft and under 5 ft 3 in.; 5 ft 3 in. and under 5 ft 4 in.; and so on.

**Table 18**   Frequency distribution of personal incomes after tax in the United Kingdom for 1972–3

| Range of incomes after tax (lower limit) | Number of incomes (thousands) |
|---|---|
| 595 | 1,808 |
| 750 | 3,004 |
| 1,000 | 2,875 |
| 1,250 | 2,743 |
| 1,500 | 2,635 |
| 1,750 | 2,103 |
| 2,000 | 2,909 |
| 2,500 | 1,138 |
| 3,000 | 719 |
| 4,000 | 183 |
| 5,000 | 95 |
| 6,000 | 79 |
| 8,000 | 20 |
| 10,000 | 6 |
| 12,000 | 2 |
| 15,000 | 1 |
| *Total* | 20,319 |

In some cases, however, it is impossible to keep the class-interval constant and at the same time to present a satisfactory frequency table. A good example is the distribution of incomes: a large class-interval would sacrifice all the detail about the smaller incomes, which form the great majority, and a small class-interval would either give an unwieldy table or involve lumping together all incomes over, say, £10,000.

Table 18, taken from *Inland Revenue Statistics 1975*, will illustrate the difficulty. The variable is income after tax, and there are sixteen classes, with seven different class-intervals, including the open group at the end. The difficulty is even greater with incomes before tax.

## Choice of class-intervals and class-limits

The student will learn from experience how to group the data in any particular case, but it is generally best to have between ten and twenty classes with equal class-intervals of convenient size. Great care should be taken to define the class-limits, i.e. the end values of each class. Thus, Table 17 might have been written in the following way:

**Table 19**  Frequency distribution
of 1000 men by height

(How *NOT* to do it)

| Height (inches) | No of men |
|---|---|
| −58 | 2 |
| 58–60 | 5 |
| 60–62 | 14 |
| 62–64 | 60 |
| 64–66 | 187 |
| 66–68 | 304 |
| 68–70 | 263 |
| 70–72 | 121 |
| 72–74 | 36 |
| 74–76 | 7 |
| 76– | 1 |
| *Total* | 1000 |

Table 19 is ambiguous because it is not clear whether a man recorded as exactly 64 inches tall should be placed in the fourth or the fifth class, whereas Table 17 makes it clear that he is placed in the fifth. The critical student may object that no man would be exactly 64 inches tall, but that any man would be either under or over 64 inches if measured accurately enough. As already remarked, however, it is not possible to measure with perfect accuracy, and in all probability the heights have been taken to the nearest $\frac{1}{4}$-inch or $\frac{1}{2}$-inch, so there will be a considerable number of men whose recorded heights correspond to class-limits. It is therefore essential to show clearly in which class they are placed.

It should be noted that if each man's height is measured correct to, say, the nearest $\frac{1}{2}$-inch, an observation of 62 inches really means '$61\frac{3}{4}$ and under $62\frac{1}{4}$', $63\frac{1}{2}$ inches means '$63\frac{1}{4}$ and under $63\frac{3}{4}$', and so on. But the class '62 and under 64' includes recorded measurements of 62, $62\frac{1}{2}$, 63 and $63\frac{1}{2}$ inches only; consequently, the correct description would be '$61\frac{3}{4}$ and under $63\frac{3}{4}$', i.e. the frequency table has overstated the class-limits by $\frac{1}{4}$-inch.

This point is worth bearing in mind, particularly in dealing with averages (see Chapter 10). Notice, incidentally, how Table 18 has avoided the problem by giving the lower limit of each interval.

### Frequency diagrams

There are various methods of representing frequency distributions graphically. In the case of a discrete variable with a small number of possible values, the simplest way is to erect vertical lines at regular intervals along the base line, with lengths proportional to the corresponding frequencies. Thus, for a graph of Table 16, points on the base line could be marked off at intervals of, say, $\frac{1}{4}$-inch or $\frac{1}{2}$-inch, representing 0, 1, 2 ... 8 children, and vertical lines erected at these points to heights, of, say, 1·79, 2·05 ... 0·01 inches, as in Fig. 19.

Alternatively, columns of equal thickness could be erected instead of lines, on the principle of the bar chart. For a discontinuous variable this method is generally preferable, and to make it still clearer the actual frequency can be written over each column if desired.

### Histograms

The above methods are unsuitable for a continuous variable, and it is best to regard the base line as an *x*-axis, on which *every* point represents a possible value of the variable. It is now impracticable to erect vertical lines or columns since the actual values of the items are generally not specified but only located in a range. In such a case the class-frequencies may be represented by means of a *histogram*.

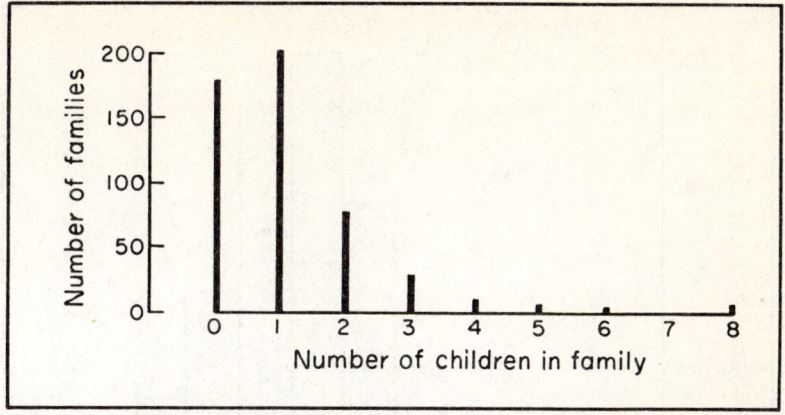

**Fig. 19** Frequency diagram representing distribution of families by number of children (see Table 16)

To construct this, points are marked off on the base line corresponding to the class-limits, so that the base line—or at least part of it—is divided into a convenient number of sections corresponding to the class-intervals and proportional to them in length. On each section is drawn a rectangle whose *area* represents the corresponding class-frequency. The result is a series of rectangles whose total area represents the total number of items observed. These rectangles comprise the histogram, which forms a continuous area provided there are no missing classes, or classes with frequency zero.

Fig. 20 shows a histogram of the height distribution given in Table 17. As the reader will see later, the general form is one that occurs frequently in statistics.

Notice that the two extreme classes, which have no definite interval in the table, have been treated in the histogram as having the same class-interval as the others, viz. 2 inches. As the numbers in these two classes are very small, little accuracy has been lost in this way. The only reasonable alternative was to omit them altogether. When the class-intervals are not all equal the problem is a little more difficult: the alternatives are either to omit the extreme classes entirely (provided the numbers are small) or to assign to them reasonable but arbitrary class-intervals.

Notice also that the ordinate in Fig. 20 is described as 'Number of men per class-interval'. The description is only legitimate when the class-intervals are equal, since it is the area, not the height, of the rectangle that represents the frequency. It comes to the same thing

**Fig. 20** Histogram representing frequency distribution of heights of 1000 men (see Table 17)

when the class-intervals are equal, but not otherwise, as the following simple illustration will demonstrate.

Fig. 21(*a*) shows part of a histogram with equal class-intervals, and Fig. 21(*b*) shows the effect of combining rectangles B and C into one. Obviously, the height of the new rectangle should be intermediate between those of B and C, and this will in fact happen if the area is

**Fig. 21(*a*)**  Histogram

**Fig. 21(*b*)**  Histogram of Fig. 21(*a*) but with rectangles B and C combined

proportional to the frequency, so that the area of the new rectangle is the sum of areas B and C. But if the histogram had been constructed so that the *height* represented the class-frequency, the height of the new rectangle would have been the sum of the heights of B and C.

## Frequency polygons

When the class-intervals are all equal, the data can be represented by a frequency polygon. As in the histogram, the base line is divided into sections corresponding to the class-intervals, but instead of the rectangles, vertical lines are erected at the middle points of the sections with heights proportional to the class-frequencies. The points at the top of successive verticals are joined up, the extreme points being joined to the base line at the mid-points of what would have been the next sections. If squared paper is used, the points can be plotted in the usual way and there is no need for the vertical lines.

The frequency polygon is now rarely used; the histogram is preferable as a rule, since it shows more clearly the relation of each class-frequency to the total.

## Frequency curves

If the quantity of data is very large, and a small class-interval is chosen so that the number of classes is quite large, it often happens that the points of the frequency polygon follow a fairly regular course, so that a smooth curve could be drawn through, or at least near, all the points. The larger the number of classes and items the nearer, as a rule, will the polygon approach a smooth curve. This *frequency curve*, as it is called, may be regarded as the limiting form of the frequency polygon or of the histogram.

It may happen, of course, that the histogram or frequency polygon shows great irregularity, in which case there is no justification for attempting to draw a curve. On the other hand, many distributions appear to follow a fairly regular law, and in some cases there are good theoretical grounds for expecting them to do so. The data can then be smoothed by means of a frequency curve, with the confidence that if further observations were taken they would still conform closely to the curve. Examples are heights and weights of men and women, marks of examination candidates, errors of observation, and many kinds of biological data.

## Common types of frequency curve

Some of the commoner forms of frequency curve are described below. Most of this section applies equally to histograms and frequency polygons, but it will be convenient to discuss distributions in terms of frequency curves.

### (a) The symmetrical 'hump-backed' distribution

This type of curve is of great importance in the theory of statistics. The distribution is symmetrical about a central maximum, the frequencies falling away to zero on either side (see Fig. 22).

**Fig. 22**   Symmetrical hump-backed frequency curve

In practice it is unusual to find a perfectly symmetrical curve, but distributions of human heights, examination marks and experimental errors approximate to this form.

### (b) The skew 'hump-backed' distribution

This is the commonest type of distribution occurring in statistical work. As before, the frequencies tail off to zero on either side of a peak, but they do so faster on one side. than on the other, as in Fig. 23; hence, it is called a *skew* or *asymmetrical* distribution. Good examples of this type are the weight distributions of men and women, and the distribution of their ages at the time of marriage.

**Fig. 23**   Skew hump-backed frequency curve

## (c) The 'J-shaped' distribution

This type exhibits a maximum at one end of the range and tails away to zero at the other (see Fig. 24). It is often an extreme form of the second type, in which the interval between the peak and one end is so small that it has been impossible to analyse this part of the distribution so as to bring out the real form of the curve. Distributions of income and wealth follow this form.

**Fig. 24**   J-shaped frequency curve

These three types do not exhaust all possible forms of distribution, but they are well-recognized types and others may generally be regarded as variants of them.

## Cumulative frequency distributions

It is sometimes desirable to use a table of cumulative frequencies rather than that of class-frequencies. Suppose, for example, that it is required to estimate from Table 17 the number of men less than 5 ft 3 in. tall. Table 17 might be rearranged as in Table 20.

Table 20 shows that there are 21 men under 62 in. and another 60 of 62 in. or over but under 64 in. Assume that these 60 are spread evenly over this range; this may not be quite accurate, but it is the simplest assumption to make. This gives 30 men under 63 in. besides the 21 under 62 in., an estimated total of 51 men under 5 ft 3 in.

In practice, this procedure would not be necessary to answer such a simple question, but it will serve as an illustration. Table 20 could, of course, have been compiled without the middle column, only the cumulative frequencies being shown and the height being given as under 58 in., under 60 in., under 62 in., and so on.

**Table 20**   Cumulative frequency distribution of 1000 men by height

(Data of Table 17)

| Height (inches) | Number of men | Cumulative number |
|---|---|---|
| Under 58 | 2 | 2 |
| 58 and under 60 | 5 | 7 |
| 60 and under 62 | 14 | 21 |
| 62 and under 64 | 60 | 81 |
| 64 and under 66 | 187 | 268 |
| 66 and under 68 | 304 | 572 |
| 68 and under 70 | 263 | 835 |
| 70 and under 72 | 121 | 956 |
| 72 and under 74 | 36 | 992 |
| 74 and under 76 | 7 | 999 |
| 76 and over | 1 | 1000 |
| Total | 1000 | — |

The summation may begin from the upper end of the range, as in the case of large incomes or estates.

These cumulative frequency tables may be represented by cumulative frequency polygons or curves, in which the height corresponds to the cumulative frequency. The cumulative frequency polygon derived from Table 20 is shown in Fig. 25. (Strictly speaking, it is not a polygon, but a series of short straight lines.)

**Fig. 25**   Cumulative frequency polygon (from Table 20)

A casual glance will show that it is much easier to draw a smooth curve through the points on this figure than in the case of the ordinary histogram or frequency polygon. This advantage is even more marked when the class-intervals are not all equal.

Such curves are known as *distribution curves* or *ogives* (see Fig. 26). Among other things, they can be used to estimate graphically the numbers of items with values less (or perhaps greater) than some specified value. Just as Table 20 was used to estimate the number of men less than 5 ft 3 in. in height, so a similar estimate could be obtained from the corresponding ogive by reading off the height corresponding to 63 in. In fact, this method would probably give a more accurate estimate than interpolation from the table.

If desired, the cumulative frequencies can be expressed as percentages of the total and the distribution curve drawn as a percentage cumulative frequency curve. Table 20 could easily be put into this form by inserting a decimal in each number in the last column, and similarly with Fig. 26, but the total is not usually such a convenient number. However, it is quite a simple task to calculate the percentages with the aid of a slide rule or a calculating machine.

One mistake frequently made by students is to plot the cumulative frequency against the mid-point, instead of the upper limit, of the last interval to be added. Thus, they would plot 268 men against 65 in., confusing the procedure with that on the original distribution. With

**Fig. 26** Distribution curve (from Table 20)

the cumulative distribution, however, this is wrong; 268 is the number of men under 66 in., not the number under 65 in. The shape of the curve is not affected; the effect of the mistake is simply to shift the whole curve half an interval (in this case one inch) to the left.

## Exercises

9.1   The data below shows the amounts of the income tax repayments due to taxpayers in sub-district Y for the year 1971–2.
  You are required to:

(*a*)  (i)  present the data in the form of a grouped frequency table; and
       (ii)  as a histogram;
(*b*)  explain the advantages of presenting data in these forms.

*Sub-district Y Tax Repayments 1971–2*

| £ | £ | £ | £ | £ | £ | £ | £ | £ | £ |
|------|-------|-------|-------|-------|-------|-------|-------|-------|-------|
| 17·35 | 6·80 | 9·05 | 14·80 | 24·70 | 23·45 | 29·90 | 9·95 | 15·90 | 22·40 |
| 16·45 | 11·50 | 17·55 | 18·60 | 9·40 | 13·35 | 26·60 | 6·65 | 10·25 | 19·00 |
| 23·95 | 18·45 | 12·70 | 19·05 | 15·15 | 18·30 | 15·16 | 12·80 | 23·40 | 5·55 |
| 8·75 | 18·80 | 20·45 | 15·80 | 22·50 | 19·10 | 14·55 | 11·15 | 16·35 | 26·60 |

*(ACA)*

9.2   Illustrate the following data by means of a histogram, a frequency polygon and a frequency curve.

| Monthly salary £ | Number of employees |
|:---:|:---:|
| 31–35 | 2 |
| 36–40 | 3 |
| 41–45 | 4 |
| 46–50 | 9 |
| 51–55 | 12 |
| 56–60 | 7 |
| 61–65 | 4 |
| 66–70 | 1 |

How does the frequency curve for this distribution differ from a curve showing skewness?                                    *(ACA)*

9.3   The following frequency distributions show the duration (reckoned in whole days) of absence from two causes of sickness in a certain transport organization:

| Duration (days) | Number of absences A | Number of absences B |
|---|---|---|
| 1 | 1 | — |
| 2 | 2 | — |
| 3 | 4 | 6 |
| 4 | 6 | 8 |
| 5 | 9 | 13 |
| 6 | 13 | 16 |
| 7 | 20 | 17 |
| 8 | 22 | 15 |
| 9 | 16 | 11 |
| 10 | 7 | 6 |
| 11 | — | 5 |
| 12 | — | 3 |

Draw the following three diagrams:

For 'A'—histogram and frequency polygon.
For 'B'—histogram and frequency curve (freehand).
For 'A' and 'B'—ogive (two curves).                    (*IoT*)

9.4   Construct histograms to illustrate the following unemployment figures.

| Duration of unemployment in weeks | Males | Females |
|---|---|---|
| One or less | 46,675 | 13,202 |
| Over   1 and up to   2 | 37,606 | 9,229 |
| Over   2 and up to   3 | 23,671 | 4,935 |
| Over   3 and up to   4 | 22,107 | 3,941 |
| Over   4 and up to   5 | 20,422 | 3,774 |
| Over   5 and up to   6 | 19,320 | 3,612 |
| Over   6 and up to   7 | 18,542 | 3,440 |
| Over   7 and up to   8 | 15,237 | 3,086 |
| Over   8 and up to   9 | 14,555 | 2,727 |
| Over   9 and up to 13 | 47,046 | 9,186 |
| Over 13 and up to 26 | 80,893 | 13,344 |
| Over 26 and up to 39 | 40,144 | 4,986 |
| Over 39 and up to 52 | 25,424 | 3,002 |
| Over 52 | 82,534 | 8,308 |
| Total | 494,176 | 86,772 |

(Source: *Employment and Productivity Gazette*)                    (*IoS*)

9.5 A survey of people waiting at various bus stops included the following 70 timings (each measured to the nearest minute) of the number of minutes the people observed had to wait for a bus:

| | | | | | | |
|---|---|---|---|---|---|---|
| 0 | 5 | 0 | 1 | 2 | 1 | 2 |
| 1 | 0 | 6 | 3 | 5 | 3 | 4 |
| 3 | 1 | 7 | 8 | 9 | 5 | 5 |
| 7 | 2 | 17 | 0 | 2 | 4 | 1 |
| 23 | 4 | 1 | 5 | 6 | 6 | 3 |
| 1 | 5 | 19 | 4 | 1 | 0 | 5 |
| 2 | 18 | 3 | 14 | 5 | 2 | 6 |
| 5 | 0 | 5 | 15 | 3 | 5 | 3 |
| 6 | 1 | 3 | 12 | 7 | 6 | 10 |
| 1 | 3 | 1 | 2 | 0 | 4 | 1 |

Choosing suitable, but not necessarily equal, class-intervals, compile a grouped frequency distribution. By cumulating your frequency distribution, estimate the missing figures in the following statement:

... per cent of the people observed waited no more than one minute for a bus, but half of them had to wait ... minutes or more and ... per cent had to wait ten minutes or more.

In addition to the 70 timings recorded there were 3 untimed cases where people gave up and went away after waiting more than ten minutes. What would the missing figures in the statement become if these 3 people were allowed for in your estimates? (*IoT*)

9.6 From the following table of income distribution, prepare a cumulative frequency curve (ogive) and explain the purpose of presenting the data in this form.

| Range of incomes £ | Number of families |
|---|---|
| 1150–1249 | 5 |
| 1250–1349 | 4 |
| 1350–1449 | 8 |
| 1450–1549 | 13 |
| 1550–1649 | 13 |
| 1650–1749 | 11 |
| 1750–1849 | 11 |
| 1850–1949 | 9 |
| 1950–2049 | 5 |
| 2050–2149 | 6 |
| 2150–2249 | 3 |
| 2250–2349 | 1 |
| 2350–2449 | 1 |
| Total | 90 |

(*ACA*)

# 10 Averages

If an examiner were asked how the boys' marks in a certain examination compared with those of the girls, he could reply in various ways. He might show the inquirer the actual marks obtained, but unless there happened to be a very striking difference between the sexes this method would be practically useless. He might produce the frequency distributions of the boys' marks and girls' marks separately, but even then a concise comparison might be difficult. Most likely, he would say something like this: 'The average score of the boys was 3·4 more (or less) than that of the girls, but the girls' marks were rather more spread out.'

The reader will see from this simple illustration that it is often desirable to go further than the compilation of a frequency distribution, and to condense the data into two or three quantities or *parameters* characterizing the distribution: first, a measure of central tendency, a typical value round which the various items are grouped, i.e. an *average*; secondly, a measure of *dispersion*, i.e. some indication of the way in which these items are scattered about the average; and possibly, although not very often, a measure of *skewness*, the tendency of items to spread out more on one side of the average than on the other. Other parameters are sometimes used in advanced work, but they are beyond the scope of this book.

This chapter will be confined to the averages; dispersion and skewness are left to Chapter 11.

## The arithmetic mean

It is well known that the average of everyday speech is obtained by adding up, say, 47 items and dividing by 47. The word 'average', however, is used in a wider sense in statistics, the particular average just mentioned being called the *Arithmetic Mean* (AM for short), or simply the *Mean*. This is the average generally used, but there are several others which, for various reasons to be discussed later, are sometimes preferable.

The formula for the mean is very simple. If $\bar{x}$ is the mean of $n$ items $x_1, x_2, \ldots x_n$,

$$\bar{x} = \frac{1}{n}\{x_1 + x_2 + \ldots + x_n\}$$

or more briefly,

$$\bar{x} = \frac{1}{n}\Sigma x \tag{10.1}$$

where the Greek letter $\Sigma$ (capital sigma) denotes summation.

If $x_1, x_2, \ldots x_n$ occur with frequencies $f_1, f_2, \ldots f_n$ respectively, the formula becomes

$$\bar{x} = \frac{1}{N}\Sigma fx \tag{10.2}$$

where $N(=\Sigma f)$ is the total frequency.

### Calculating the mean

If there are only a few items, it is easy to add them together and divide by their number. If there are many, however, it is better to form the frequency distribution, grouping them if necessary into a convenient number of classes and treating each class as if all its items had the same value, viz. the mid-value of the class-interval. The error involved in this method will be small unless there are large class-frequencies at the extremes.

*Example* 10.1

The scores of 145 competitors in a golf tournament ranged from 66 to 89 as shown in the table below. Find the mean score.

| Score | Frequency | Score | Frequency | Score | Frequency |
|-------|-----------|-------|-----------|-------|-----------|
| 66 | 1 | 74 | 4 | 82 | 5 |
| 67 | – | 75 | 15 | 83 | 7 |
| 68 | – | 76 | 13 | 84 | 5 |
| 69 | 4 | 77 | 12 | 85 | 7 |
| 70 | 4 | 78 | 7 | 86 | 1 |
| 71 | 6 | 79 | 17 | 87 | 1 |
| 72 | 7 | 80 | 11 | 88 | 1 |
| 73 | 9 | 81 | 7 | 89 | 1 |
| | | | | Total | 145 |

If the scores are grouped in class-intervals of 5, with central values 65, 70, etc., the distribution can be re-rewritten as follows:

| Score | Central value (x) | Frequency (f) | Product (fx) |
|---|---|---|---|
| 63–67 | 65 | 1 | 65 |
| 68–72 | 70 | 21 | 1,470 |
| 73–77 | 75 | 53 | 3,975 |
| 78–82 | 80 | 47 | 3,760 |
| 83–87 | 85 | 21 | 1,785 |
| 88–92 | 90 | 2 | 180 |
| Total | | 145 | 11,235 |

$$\text{Mean} = \frac{11,235}{145} = 77 \cdot 5 \text{ approximately.}$$

## Short-cut methods

The student may wonder whether all this arithmetic is really necessary. The working can, in fact, be shortened considerably by (i) estimating the mean and taking an arbitrary origin somewhere near it, and (ii) taking the class-interval as the unit. Thus a score of 75 might be taken as an origin, or working mean, and differences from it measured in units of 5, giving the following table:

| Central value (x) | Deviation in units (d) | Frequency (f) | Product (fd) |
|---|---|---|---|
| 65 | −2 | 1 | − 2 |
| 70 | −1 | 21 | −21 |
| 75 | — | 53 | — |
| 80 | +1 | 47 | +47 |
| 85 | +2 | 21 | +42 |
| 90 | +3 | 2 | + 6 |
| Total | | 145 | −23 + 95 = +72 |

This gives $+72/145$ for the mean value of $d$, i.e. for the deviation of the mean score from 75, measured in class-intervals of 5. Hence the mean score is

$$75 + \left(\frac{72}{145} \times 5\right) = 77 \cdot 5, \text{ as before.}$$

There are several pitfalls in these short-cut methods, and the student must be on his guard until practice has made perfect. He must watch his signs carefully, i.e. the signs of the deviations, the products *fd* and their algebraic sum; it is best to sum the negative and the positive products separately. Then he must remember to multiply the mean value of *d* by the class-interval before adding it to the working mean. He may wonder whether this method is really worth while or whether it is, like many other short cuts, 'the longest way round'. When he comes to the following chapter, however, he will realize that it is even more useful in the calculation of the standard deviation.

### The weighted mean

One advantage of the arithmetic mean is that if the means of several sets of items and the number of items in each set are known, the mean of all sets combined can easily be calculated. For instance, suppose the mean age of 20 boys in a class is 14 years and that of 10 girls is 15 years. Then the total of the boys' ages is 280 years and that of the girls' ages 150 years, giving a total for the whole class of 430 years, so the mean age of the class is 430/30 years, or 14 years 4 months. And, in general, if $f_1$ items have a mean $m_1$, $f_2$ a mean $m_2$, and so on, the mean of all the items is

$$\frac{f_1 m_1 + f_2 m_2 + \ldots}{f_1 + f_2 + \ldots} \text{ or } \frac{\Sigma fm}{\Sigma f} \tag{10.3}$$

This leads to the *weighted mean*, as opposed to the simple or unweighted mean of a number of items. Suppose in the previous example the average age of the boys was known to be 14 years and that of the girls 15 years, but it was not known how many boys or how many girls there were. It would be necessary to assume that their numbers were nearly equal, to take the simple mean of 14 and 15 and to say that the average age of the class was (as nearly as could be judged) 14 years 6 months. But if it was known that there were about twice as many boys as girls, their averages could be 'weighted' with factors 2 and 1 respectively, giving the average age of the class as

$$\frac{14 \times 2 + 15 \times 1}{2 + 1} = 14\tfrac{1}{3} \text{ years, i.e. 14 years 4 months.}$$

It is evident that if the boys outnumber the girls in the ratio 2:1 exactly, this method gives the correct answer. If the ratio is not exact, but approximately correct, the result will not be far out, as the student should verify for himself by taking, say, 19 boys and 11 girls, or 23 boys and 10 girls.

In general, the weighted mean of $n$ items $x_1, x_2 \ldots x_n$, with weights $w_1, w_2 \ldots w_n$, is

$$\frac{w_1x_1 + w_2x_2 + \ldots + w_nx_n}{w_1 + w_2 + \ldots + w_n} = \frac{\Sigma wx}{\Sigma w} \tag{10.4}$$

Compare this formula with (10.3). Evidently the mean obtained from a frequency distribution is a weighted mean in which the weights are the class-frequencies.

Weighted means are used chiefly in connection with index numbers (see Chapter 17), but they are sometimes used to combine several observations or averages as in the following example.

*Example* 10.2

Three men, $A$, $B$ and $C$, conduct an experiment to estimate a certain quantity and obtain values of $11 \cdot 27$, $11 \cdot 41$ and $11 \cdot 36$. On account of their differences in skill and experience, their observations are allotted weights of 5, 3 and 2 respectively (in everyday speech, $A$'s estimate carries more weight than the others), and the weighted mean is taken as the best possible estimate of the item concerned.

Taking differences from $11 \cdot 3$, the weighted mean is

$$\frac{5(-0 \cdot 03) + 3(0 \cdot 11) + 2(0 \cdot 06)}{5 + 3 + 2} = 0 \cdot 03$$

Hence the quantity is estimated as $11 \cdot 33$.

## The median

Sometimes it is desirable to use a simpler average than the mean, one that is more easily calculated or for other reasons more suitable for the purpose in hand. One such average is the *Median*, defined as the central value of the variable. Thus if 37 items are arrayed in order of magnitude, the median is the value of the middle one, i.e. of the 19th. If there are 38 items, there are two that have an equal claim to be central, the 19th and 20th, so the median is taken as the average of these two. To take a simple example, if five men shoot and score 48, 55, 59, 67 and 70, the median score is 59; but if a sixth man scores 62, the median is $\frac{1}{2}$ (59 + 62), i.e. $60\frac{1}{2}$.

When the data are grouped, matters are not quite so simple, as the median can only be estimated, not exactly located. The following example will show the method to be employed and should be studied carefully.

*Example* 10.3

Find the median height of the frequency distribution shown in Table 17.

The cumulative frequency table is first constructed as in Table 20 until half the total frequency has been reached, which in this case occurs in the sixth group. It is unnecessary to proceed further with the table unless the median is being found graphically (see below).

| Height (inches) | Number of men | Cumulative number of men |
|---|---|---|
| Under 58 | 2 | 2 |
| 58 and under 60 | 5 | 7 |
| 60 and under 62 | 14 | 21 |
| 62 and under 64 | 60 | 81 |
| 64 and under 66 | 187 | 268 |
| 66 and under 68 | 304 | 572 |

Obviously the median is in the group 66–68 in., and common sense suggests that it is nearer 68 in. than 66 in. The best assumption to make is that the heights of the 304 men in this group are evenly spread over the interval. Now there are 500 men below the median ($M$, say) and 268 men below 66 in., so on the above assumption the distance ($M - 66$) in. is $(500 - 268)/304$ of the class-interval which is 2 inches.

$$\text{Hence } M = 66 + \frac{500 - 268}{304} \times 2$$

$$= 67 \cdot 53 \text{ in. approximately.}$$

The assumption made here is not strictly correct, but the error involved is generally small.

It is quite easy to estimate the median graphically and, where items are not distributed evenly over the class containing it, this method often gives a more accurate result. Construct the ogive and read off the value of the variable corresponding to the point whose height is half the total frequency. Fig. 27 shows the ogive for the data of Example 10.3, but continuing after the sixth group. The value of $x$ corresponding to $y = 500$ is $67 \cdot 5$ in. as nearly as can be judged.

**Fig. 27** Estimation of the median of a grouped distribution (from Table 20)

## The mode

A third very simple average is the *Mode*, the value of the variable occurring most often, corresponding to the highest point of the frequency curve. The mode is difficult to determine accurately from grouped data unless the number of data is very large. With a discrete variable it can usually be found at once, although if the frequencies are fairly small, peculiar results may occur through sampling errors (see Chapter 13). If the variable is continuous, it is necessary to estimate what value would have the greatest frequency if there were enough data. In the case of Table 17, for instance, it is not at all clear whether, if each man's height were measured correct to 1/10 in., the mode would be 67·5 in., 67 in. or 68 in., or any other particular value. The best way to estimate the mode would be to fit a smooth curve to the data and take the variable corresponding to the peak.

## Relation between the mean, median and mode

When the frequency distribution is perfectly symmetrical, these three averages all coincide. If, however, the distribution is skew, they may differ appreciably. For most frequency curves of the 'cocked-hat' type, the median lies between the mean and the mode, the distances being such that:

Mean − mode = 3(mean − median) approximately.

In difficult cases, this relation may be used to estimate the mode of a grouped distribution of the 'cocked-hat' type. It is not exact, but it works out remarkably well in practice. It is helpful to remember that when the longer tail of the distribution lies on the left, the mean, median and mode are in dictionary order and that the median lies nearer to the mean than to the mode—again, as in the dictionary. When the longer tail is on the right the order is, of course, reversed.

Fig. 28 illustrates these three averages with reference to a frequency curve. The verticals $M_1$, $M_2$ and $M_3$ correspond respectively to the mean, the median and the mode. The latter gives the highest point of the curve, $M_1$ passes through the centre of gravity of the area, and $M_2$ divides the area into two equal parts.

**Fig. 28**  Frequency curve showing mean, median and mode

## Relative advantages and disadvantages

Before proceeding to consider other possible averages, it will be useful to examine the relative merits of the three already discussed.

The mean possesses several great advantages: it is easily understood and, as a rule, easily calculated, and it takes every item into account equally. The median is also easy to understand and to calculate; indeed, it is often much easier to calculate than the mean, but it depends only on the middle items or groups, as does the mode, and is not affected by any change in the extreme items. Both the mean and the median are uniquely determined by the items, except for small errors in the case of grouped data, whereas the mode is generally difficult to estimate. There may indeed be more than one mode.

The chief advantage of the mean, however, is that it lends itself to algebraic treatment or to further calculations. It has been shown how, when two or more distributions are combined into one, their means can be weighted to find the mean of the whole. This cannot be done with the median or the mode. However, one must remember that if a distribution is clearly not homogeneous, but a mixture of two or more groups with very different distributions (e.g. firms paying very different salaries), no average means very much.

The mean also has the advantage that the total value of the population can be worked back from the number of items and their mean value. For instance, if the average wage of 1000 men is £58 a week, the weekly wage bill is £58,000; this could not be deduced from knowledge of the mode or median. There are, of course, cases where the total is of little interest: if the mean age of a class of 30 boys is 10 years, it is not very useful to know that their total is 300 years. But if their mean pocket money is 35p a week, it is useful to know that their total pocket money is £10.50, particularly to the master responsible for distributing it to them.

It must not be supposed, however, that all the advantages are with the mean. If the population contains a few exceptional cases, they may distort the mean unduly; for example, it would be misleading to quote the mean income of a small community if it happened to include a millionaire. Again, the median may be chosen for economic reasons: if a batch of 25 lamps is being tested by the time taken to burn out, it will save both time and material if the time taken for the 13th lamp to burn out is taken as the average life of a lamp. Again, where the extreme groups of a distribution have no definite range, it is difficult to determine the mean accurately. The median is unaffected by this.

These three averages having been discussed at some length, it only remains to consider briefly two of minor importance.

## The geometric mean

The *Geometric Mean* (GM) of $n$ observations $x_1, x_2, \ldots x_n$ is given by the formula

$$g = \sqrt[n]{(x_1 x_2 \ldots x_n)} \qquad (10.5)$$

It is generally calculated by taking logarithms, since

$$\log g = \frac{1}{n} \Sigma \log x \qquad (10.6)$$

*Example* 10.4
Find the geometric mean of the numbers 104, 119, 86, 94, 155 and 77.

| Number | Log |
|--------|-----|
| 104 | 2·0170 |
| 119 | 2·0755 |
| 86 | 1·9345 |
| 94 | 1·9731 |
| 155 | 2·1903 |
| 77 | 1·8865 |
| | 6)12·0769 |
| GM = 103·0 | 2·0128 |

The student should find the AM and compare it with the GM.

The GM is used almost entirely in the calculation of index numbers (see Chapter 17), and should only be used when all the items are positive. It has the effect of reducing the influence of large items and increasing that of small ones. It is easily proved (the proof is found in many algebra books) that the GM is always less than the AM.

### The harmonic mean

The *Harmonic Mean* (HM) of $n$ observations $x_1, x_2, \ldots x_n$ is given by the formula

$$\frac{1}{h} = \frac{1}{n} \sum \frac{1}{x} \tag{10.7}$$

Thus, as with the GM, finding the HM reduces to finding the AM with a simple transformation of the variable. For the GM we take logarithms, for the HM we take reciprocals.

*Example* 10.5

Find the harmonic mean of 20 and 30.

$$\frac{1}{h} = \frac{1}{2} \left( \frac{1}{20} + \frac{1}{30} \right) = \frac{5}{120} = \frac{1}{24}$$

$$\therefore h = 24.$$

The usual illustration is average speed over a return journey. Thus, a cyclist might cover a distance $d$ miles at an average speed of 20 mph and the same distance back at an average speed of 30 mph. The total time taken is $d/20 + d/30$ hours. It must also be $2d/h$ hours, where $h$ mph is the average speed over the double journey. Example 10.5 shows that $h = 24$, the HM of 20 and 30.

In practice, the HM is very seldom used. It is only mentioned here because examiners sometimes set questions on it.

## Exercises

10.1 The following figures represent the weekly expenditure on fruit and vegetables by each of 60 families in a sample.

| £ | £ | £ | £ | £ | £ |
|------|------|------|------|------|------|
| 0·98 | 0·71 | 1·05 | 0·79 | 1·53 | 1·38 |
| 1·24 | 1·76 | 0·78 | 0·89 | 0·91 | 1·03 |
| 0·78 | 0·84 | 0·91 | 0·93 | 1·12 | 1·25 |
| 1·26 | 0·87 | 0·98 | 0·74 | 0·51 | 0·38 |
| 0·33 | 1·19 | 1·48 | 1·62 | 0·41 | 0·48 |
| 0·69 | 0·88 | 1·08 | 1·03 | 1·78 | 0·69 |
| 0·51 | 0·58 | 1·28 | 1·51 | 1·33 | 0·66 |
| 0·74 | 0·51 | 1·58 | 1·17 | 1·79 | 0·64 |
| 1·61 | 0·48 | 1·16 | 1·26 | 1·44 | 1·83 |
| 0·46 | 0·36 | 0·21 | 1·96 | 2·13 | 0·87 |

You are required to

(a) arrange the figures in the form of a grouped frequency distribution;
(b) calculate the average weekly family expenditure on fruit and vegetables, using the short-cut method;
(c) plot the figures in the form of a histogram.

*(based on ICMA)*

10.2 Calculate the mean strength of the following frequency distribution of lots of cotton yarn:

Strength of lots of cotton yarn

| Pounds per square inch | Number of lots |
|------|------|
| 70– 74 | 1 |
| 75– 79 | 6 |
| 80– 84 | 17 |
| 85– 89 | 29 |
| 90– 94 | 20 |
| 95– 99 | 17 |
| 100–104 | 13 |
| 105–109 | 10 |
| 110–114 | 6 |
| 115–119 | 3 |
| 120–124 | 2 |
| 125–129 | 1 |

State what your answer indicates and compare the mean with the median as a statistical average. (*ACA*)

10.3 The following figures give the daily flows through a sewage plant, in thousands of gallons, on 35 consecutive days. Arrange them in a frequency distribution and calculate *three* measures of central tendency.

118, 110, 152, 168, 160, 168, 119, 92, 175, 152, 168, 144, 144, 184, 152, 168, 168, 104, 137, 144, 170, 124, 114, 202, 187, 192, 221, 192, 168, 175, 175, 181, 211, 168, 211.

(*IoS*)

10.4 The arrival time of a certain train is recorded on 25 separate occasions. It is late on each occasion, the number of minutes it is late being given in Table A below. After steps are taken to improve its timekeeping, the time lost on the next 25 occasions is as in Table B.

Calculate the arithmetic mean and the median for each group of observations. Do your results enable you to judge whether the train's timekeeping has improved? Find an alternative way, by further calculation or graphically, of illustrating the differences between the two groups of figures.

| Table A | | | Table B | | |
|---|---|---|---|---|---|
| 2 | 33 | 13 | 5 | 25 | 3 |
| 1 | 18 | 28 | 14 | 17 | 23 |
| 8 | 13 | 31 | 7 | 2 | 15 |
| 6 | 8 | 2 | 10 | 9 | 21 |
| 23 | 1 | 10 | 22 | 20 | 17 |
| 14 | 4 | 27 | 17 | 9 | 6 |
| 38 | 1 | 12 | 18 | 5 | 8 |
| 29 | 22 | | 3 | 23 | |
| 18 | 32 | | 19 | 17 | |

(*IoT*)

10.5 Calculate the median of the following frequency distribution and also estimate the percentage of all the cars travelling at 33 miles per hour or more.

| Miles per hour | Number of cars (*f*) |
|---|---|
| 15·0–19·9 | 4 |
| 20·0–24·9 | 16 |
| 25·0–29·9 | 40 |
| 30·0–34·9 | 18 |
| 35·0–39·9 | 6 |
| 40·0 and over | 4 |

What does the median value of this distribution indicate, and what are the advantages of the median as a statistical average?    (*ACA*)

10.6

Mangolds: estimated yield per acre

| County | Tons | County | Tons |
|---|---|---|---|
| Bedfordshire | 32·7 | Derbyshire | 20·9 |
| Berkshire | 24·4 | Devon | 27·1 |
| Buckinghamshire | 31·5 | Dorset | 34·5 |
| Cheshire | 18·5 | Durham | 20·9 |
| Cornwall | 26·0 | Essex | 24·2 |
| Cumberland | 19·0 | Kent | 25·6 |

(Source: *Agricultural Statistics* 1963/4)

Calculate the mean, median and mode of the above data. Which average do you here consider to be the most useful and why?

(*IoS*)

10.7    Define the weighted arithmetic mean of a set of observations $x_1, x_2, \ldots x_n$ with weights $W_1, W_2, W_3 \ldots W_n$.
The following data refer to a group of observations:

| Commodity | Weight | Observations |
|---|---|---|
| a | 4 | 113 |
| b | 5 | 140 |
| c | 3 | 120 |
| d | $W_d$ | 130 |
| e | $W_e$ | 108 |
| f | $W_f$ | 105 |

Given that the weighted arithmetic mean of the six observations is 117·68 and that the sum of the six weights is 25, calculate the weighted arithmetic mean of commodities (i) *a, b,* and *c,* (ii) *d, e,* and *f.*

(*IoS*)

10.8 Define in your own words the arithmetic mean, the geometric mean, and the harmonic mean. Calculate each of these measures using the following figures:

3, 8, 12, 6, 5, 17, 9, 15.

Give an example to show when you would use the harmonic mean in preference to the arithmetic mean, and explain why you would do so.

(*IoT*)

# 11 Dispersion and skewness

An average is not in itself an adequate summary of a frequency distribution; it is also necessary to have some idea of the way individual items are scattered about that average. Average income, for example, does not mean very much if great wealth and poverty exist side by side in a community. Also, as will appear later (Chapter 13), it is often necessary to measure the variability of a distribution in order to estimate the accuracy of a test result or to assess its significance.

## The range

A rough and ready measure of dispersion is the difference between the highest and lowest values observed, called the *range*. It is not very satisfactory because it depends solely on the two extreme values and may be very misleading if there are one or two abnormal items, e.g. if one of a group of examination candidates is a genius. Moreover, since exceptional items are more likely to occur when the number of items is large, the range tends to increase as the number of items increases. In some cases, e.g. where there are large open groups at the ends like '65 years and over', it is impossible to estimate the range. There are, however, circumstances in which the average range of small groups of items provides a good estimate of the dispersion (see Chapter 16).

## Values of position

A good way of showing how items are distributed is to find the values that divide the distribution into a number of equal parts when the data are arranged in order of magnitude. For example, the nine *deciles* divide the distribution into ten equal parts, one-tenth of the items falling below the first decile, two-tenths below the second, and so on.

*Percentiles* go further and divide the distribution into 100 equal parts when the data are very numerous. They are greatly favoured by psychologists, and the 'percentile rank' is often used when the percentile itself, as a numerical value, does not exist. Thus boys could be assessed in respect of capacity for leadership, say, and a percentile rank of 85 would mean that a boy excelled 85 per cent of his fellows in that respect.

## Quartiles

The most commonly used values of position are the *quartiles*, which divide the data into four equal parts. The second quartile has already been introduced as the median, by which name it is always known. The other two are called the lower and upper quartiles and are generally denoted by $Q_1$ and $Q_3$ respectively.

**Fig. 29**  Array of forty-one items showing positions of quartiles and deciles

Fig. 29 shows 41 items represented graphically in the form of an array, i.e. arranged in order of magnitude. It also shows the deciles and quartiles. The lower quartile is the value of the 11th item, since there are 10 below and 30 above; similarly, the upper quartile is the value of the 31st item. The deciles have been taken as the values of the 5th, 9th ... 37th items; the median is shown as $M$.

Had there been only 40 items, the lower quartile could have been taken as the average of the 10th and 11th, as there would then be 10 items below it and 30 above; similarly with the upper quartile. When there are 42 or 43 items, the position is not so clear, as no value will divide the distribution exactly in the ratio $1:3$. Probably the best rule in such cases is to take the quartiles as corresponding to the positions $(n + 1)/4$ and $3(n + 1)/4$, where $n$ is the number of items, rounding these quantities to the nearest integer where necessary. Thus, if $n = 42$, the quartiles will be the values of the 11th and 32nd items; if $n = 43$, the values of the 11th and 33rd items.

Statisticians are not unanimous about this, but if there are many items it makes no appreciable difference whether one value or the next or the average of the two is taken as the quartile; if there are few items, the error involved is small compared with errors of sampling (see Chapter 13).

## Quartiles of a grouped distribution

When the data are grouped, the quartiles cannot normally be determined accurately; they can only be estimated as the median is estimated (see page 98), either by interpolation or by means of the ogive.

*Example* 11.1
To find the quartiles of the height distribution given in Table 17.

The method is the same as in Example 10.3, where the object was to find the median. The cumulative frequency table, shown completed this time as in Table 16, is as follows:

| Height (inches) · | Number of men | Cumulative number of men |
|---|---|---|
| Under 58 | 2 | 2 |
| 58 and under 60 | 5 | 7 |
| 60 and under 62 | 14 | 21 |
| 62 and under 64 | 60 | 81 |
| 64 and under 66 | 187 | 268 |
| 66 and under 68 | 304 | 572 |
| 68 and under 70 | 263 | 835 |
| 70 and under 72 | 121 | 956 |
| 72 and under 74 | 36 | 992 |
| 74 and under 76 | 7 | 999 |
| 76 and over | 1 | 1000 |

Evidently $Q_1$ is in the range 64 to 66 in. and, as there are 250 items below $Q_1$,

$$Q_1 = 64 + \frac{250 - 81}{187} \times 2 = 65 \cdot 81 \text{ in.}$$

Similarly,

$$Q_3 = 68 + \frac{750 - 572}{263} \times 2 = 69 \cdot 35 \text{ in.}$$

The interpolation method shown above assumes that the items in any particular class-interval are evenly distributed. No such assumption is made in the graphical method. The ogive for the above table is shown in Fig. 30.

**Fig. 30**  Distribution curve showing the estimation of the quartiles (from Table 20)

As nearly as can be judged from the graph,

$$Q_1 = 65 \cdot 8 \text{ in.}, \quad Q_3 = 69 \cdot 2 \text{ in.}$$

The median has already been estimated (Example 10.3) as $67 \cdot 5$ in., so these results could be expressed in the following form:

25% of the men are under 65·8 in.
50%   ,,    67·5 in.
75%   ,,    69·2 in.

or alternatively,

25% of the men are under 65·8 in.
25% of the men are between 65·8 and 67·5 in.
25% of the men are between 67·5 and 69·2 in.

and

25% of the men are over 69·2 in.

Quartiles, deciles and similar quantities—quantiles, as they are sometimes called—are commonly used in connection with incomes, wages, and social inquiries generally.

## The quartile deviation

The quantity $\frac{1}{2}(Q_3 - Q_1)$ is called the *quartile deviation* or, less often, the *semi-interquartile range*. There is no recognized symbol for it, but it is commonly abbreviated to QD. When the distribution is symmetrical, the QD is equal to the difference between the median and either quartile.

Using the data of Example 11.1,

$$QD = \tfrac{1}{2}(69\cdot2 - 65\cdot8)\ \text{in.} = 1\cdot7\ \text{in.}$$

## The mean deviation

The *Mean Deviation* differs from the range and the QD in that it takes every item into account. It is the arithmetic mean of all deviations from the mean (or, alternatively, from the median), counting all such deviations as positive. The proviso is necessary because the algebraic sum of all deviations from the mean is zero.

Consider the numbers 19, 22, 36, 27, 16. Their mean is 24, and their deviations from the mean are $-5$, $-2$, $+12$, $+3$ and $-8$, which add up to zero, as they must. Ignoring the signs, the Md is

$$\tfrac{1}{5}(5 + 2 + 12 + 3 + 8) = 6\cdot0.$$

In practice, such a small number of items would not call for a measure of dispersion, but the example is given for simplicity.

At first sight this seems an ideal method of measuring dispersion. What could be more natural, having found the mean, than to take the average difference from it? Unfortunately, the MD has certain serious disadvantages and is therefore seldom used. In particular, it does not lend itself to further calculation or algebraic treatment as the mean does for averages. Difficulties also arise in computing the MD of a grouped frequency distribution.

Sometimes the MD is measured from the median. It is easily seen that the mean deviation is minimum when measured from this point, for if the origin of measurement is moved from the median to some other point, some deviations will be reduced, but at least as many will be increased each by the same amount.

## The standard deviation

None of the three possible measures of dispersion described above (the range, the QD and the MD) lends itself to mathematical treatment. Here the *Standard Deviation* has the advantage over them, as will be seen. That is the chief justification for what at first sight seems a clumsy and troublesome quantity to compute. Instead of the deviations themselves being averaged, as in the MD, the deviations from the mean are squared, then the sum of the squares is divided by the number of items, and the square root of the quotient is the standard deviation.

The SD is denoted by the Greek letter $\sigma$ (small sigma, not to be confused with the summation sign $\Sigma$, which is capital sigma), and $\sigma^2$ is called the *variance* of the distribution. If there are $n$ items $x_1, x_2, \ldots x_n$, with a mean $\bar{x}$, then by definition

$$\sigma^2 = \frac{1}{n}\{(x_1 - \bar{x})^2 + (x_2 - \bar{x})^2 + \ldots + (x_n - \bar{x})^2\}$$

$$= \frac{1}{n}\Sigma(x - \bar{x})^2 \tag{11.1}$$

To take the trivial example of the previous section, the SD of the numbers 19, 22, 36, 27 and 16 is given by

$$\sigma^2 = \tfrac{1}{5}\{(-5)^2 + (-2)^2 + 12^2 + 3^2 + (-8)^2\}$$
$$= \tfrac{1}{5}(25 + 4 + 144 + 9 + 64)$$
$$= 49\cdot2$$
$$\therefore \ \sigma = \sqrt{49\cdot2} = 7\cdot01 \text{ approximately.}$$

Normally, the SD would not be calculated for only five items. It has been done here to explain the process as clearly and simply as possible, because it is essential that this very important measure and the method of calculating it should be thoroughly understood.

In the above example the mean was a whole number and the sum of squares was easily calculated. Generally, however, the mean is a fraction or decimal, and to simplify, the arithmetic deviations are measured from the origin or some value near the mean, and an adjustment made to correct this. If the working mean is taken as origin, the formula is

$$\sigma^2 = \frac{1}{n}\Sigma x^2 - \bar{x}^2$$

The proof of this formula is as follows.
By (11.1) above

$$n\sigma^2 = \Sigma(x - \bar{x})^2 = (x_1 - \bar{x})^2 + (x_2 - \bar{x})^2 + \ldots + (x_n - \bar{x})^2$$

Expanding the contents of the brackets and noting that $\bar{x}$ is a constant,

$$n\sigma^2 = \Sigma x^2 - 2(\Sigma x)\bar{x} + n\bar{x}^2$$

But

$$\Sigma x = n\bar{x}$$
$$\therefore n\sigma^2 = \Sigma x^2 - 2n\bar{x}^2 + n\bar{x}^2$$
$$= \Sigma x^2 - n\bar{x}^2$$

and

$$\sigma^2 = \frac{1}{n}\Sigma x^2 - \bar{x}^2 \tag{11.2}$$

Substituting $\frac{1}{n}\Sigma x$ for $\bar{x}$ gives the alternative form

$$\sigma^2 = \frac{1}{n}\left\{\Sigma x^2 - \frac{1}{n}(\Sigma x)^2\right\} \tag{11.3}$$

where

$$(\Sigma x)^2 = (x_1 + x_2 + \ldots + x_n)^2$$

This form is usually more convenient for numerical work for, even with large numbers, $\Sigma x$ and $\Sigma x^2$ are easily computed with a calculating machine. Incidentally, formula (11.2) shows that the sum of squares $\Sigma x^2$ is least when the deviations are measured from the mean.

If $x_1$ occurs $f_1$ times, $x_2$ $f_2$ times, and so on, the formulae become

$$\sigma^2 = \frac{1}{N}\Sigma f x^2 - \bar{x}^2 \tag{11.2a}$$

and

$$\sigma^2 = \frac{1}{N}\left\{\Sigma f x^2 - \frac{1}{N}(\Sigma f x)^2\right\} \tag{11.3a}$$

where $N(= \Sigma f)$ is the total number of items.

*Example* 11.2

To find the standard deviation of the following distribution of motor drivers having 0, 1, 2 etc. accidents over a period of years.

| No of accidents | No of drivers |
|:---:|:---:|
| 0 | 45 |
| 1 | 36 |
| 2 | 40 |
| 3 | 19 |
| 4 | 12 |
| 5 | 8 |
| 6 | 3 |
| 7 | 2 |
| 8 | 1 |
| Total | 166 |

There is no point here in taking a trial mean, or even in calculating the mean. The calculation is as follows

| $x$ | $f$ | $fx$ | $fx^2$ |
|:---:|:---:|:---:|:---:|
| 0 | 45 | — | — |
| 1 | 36 | 36 | 36 |
| 2 | 40 | 80 | 160 |
| 3 | 19 | 57 | 171 |
| 4 | 12 | 48 | 192 |
| 5 | 8 | 40 | 200 |
| 6 | 3 | 18 | 108 |
| 7 | 2 | 14 | 98 |
| 8 | 1 | 8 | 64 |
| Total | 166 | 301 | 1029 |

The 'crude sum of squares'

$$= \Sigma fx^2 = 1029$$

and the 'correction term'

$$= \frac{1}{N}(\Sigma fx)^2 = \frac{301^2}{166} = 545 \cdot 8$$

∴ 'Corrected sum of squares'

$$= 1029 - 545 \cdot 8 = 483 \cdot 2$$

$$\therefore \sigma^2 = \frac{483 \cdot 2}{166} = 2 \cdot 91$$

$$\therefore \sigma = 1 \cdot 71 \text{ approximately.}$$

## The standard deviation of a grouped distribution

With a grouped distribution it is generally best to take the mid-point of one group as a working mean and measure deviations from it with the class-interval as unit or, if the class-intervals are not all equal, with some other convenient unit.

*Example* 11.3

To find the SD of the distribution of golf scores given in Example 10.1, grouped in intervals of 5 (see pages 94 and 95).

| Central value | x | f | fx | fx² |
|---|---|---|---|---|
| 65 | −2 | 1 | − 2 | 4 |
| 70 | −1 | 21 | −21 | 21 |
| 75 | — | 53 | — | — |
| 80 | +1 | 47 | +47 | 47 |
| 85 | +2 | 21 | +42 | 84 |
| 90 | +3 | 2 | + 6 | 18 |
| *Total* | | 145 | +72 | 174 |

Crude sum of squares

$$= 174$$

Correction term

$$= \frac{72^2}{145} = 35 \cdot 75$$

Corrected sum of squares

$$= 138 \cdot 25$$

$$\therefore \sigma = \sqrt{\frac{138 \cdot 25}{145}} = 0 \cdot 976 \text{ units}$$

Since deviations have been measured in units of 5 strokes,

$$SD = 0 \cdot 976 \times 5 = 4 \cdot 88 \text{ strokes}$$

The true SD, calculated from the original data, is 4·54 strokes. Considering that the grouping is so broad and that all but three items are in four groups only, this difference is to be expected. It can, however, be reduced by what is called 'Sheppard's correction for grouping'. This consists in deducting $\frac{1}{12}h^2$ from the mean square before taking the square root, $h$ being the class-interval. In the above example, taking the class-interval as unit,

$$\sigma^2 = \frac{138 \cdot 25}{145} - \frac{1}{12} = 0 \cdot 870$$

$$\therefore \ \sigma = 0 \cdot 933 \text{ units}$$
$$= 4 \cdot 66 \text{ strokes}$$

This correction applies only to distributions of the hump-backed type and not, for example, to J-shaped distributions. The justification for the value $1/12$ is beyond the scope of this book.

## Combining two or more distributions

If two or more sets of similar observations are pooled, the mean and SD of each set being known, as well as the number of observations, then we can calculate the mean and SD of the aggregate. This is seldom required, except in examination questions, but it may be useful to show how it can be done by means of a simple numerical example.

*Example* 11.4

In a test, Class A, with 30 boys, has a mean of 62 marks and an SD of 11; Class B, with 25 boys, has a mean of 53 and an SD of 8. Find the mean and SD of the two classes combined.

To simplify calculation, let us take differences from 50, making the means of A and B 12 and 3. (The SDs are unaffected.) The combined mean is

$$\frac{(12 \times 30) + (3 \times 25)}{30 + 25} = \frac{435}{55} = 7 \cdot 91,$$

or adding back the 50, 57·91 marks.

To find the SD, we must calculate first the crude sums of squares (of deviations from 50), then the corrected sum of squares. Notice that formula (11.2) can be put into the form

$$\Sigma x^2 = n(\sigma^2 + \bar{x}^2), \text{ or } n\sigma^2 + \frac{(\Sigma x)^2}{n}$$

So, crude SS for A $= 30 \, (11^2 + 12^2)$     $= 7950,$
crude SS for B $= 25 \, (8^2 + 3^2)$       $= 1825,$
crude SS for A and B together       $= 9775$

$\therefore$ corrected SS for A and B together       $= 9775 - \dfrac{435^2}{55} = 6335$

$$\therefore \sigma^2 = \frac{6335}{55} = 115 \cdot 18,$$

$$\sigma = 10 \cdot 73$$

## Comparison of the various measures of dispersion

Apart from the range, which for reasons already stated is an unreliable measure of dispersion, the easiest measure to compute and to interpret is the QD. Unfortunately it does not lend itself easily to mathematical treatment, but it is sometimes preferable to the MD or SD for a grouped distribution when the extreme groups are large or have no definite range. The MD is also easy to understand and easy to calculate for a moderate number of items, but grouped data may be troublesome.

Only the SD is suitable for algebraic manipulation. For instance, as we have seen, two or more distributions of similar data with known means and standard deviations can be combined into one distribution whose mean and SD can be calculated. This is not possible with, say, the median and the QD. The SD can also be used in precise mathematical tests of significance (see Chapter 13), and it can be proved that for the hump-backed type of distribution it is the most efficient measure of dispersion, i.e. it can be estimated from a sample more accurately than any other. It is difficult to calculate, however, when a grouped distribution has unequal class-intervals or extreme groups with no definite range or mean.

There is no definite relation between the QD, MD and SD, but in a hump-backed distribution that is either symmetrical or only moderately skew, the QD is generally about $\frac{2}{3}\sigma$ and the MD about $\frac{4}{5}\sigma$. It is evident that the same measurement must be used to compare several distributions, e.g. it is no use comparing the QD of one with the SD of another.

## Coefficient of variation, etc.

It is sometimes less important to measure the absolute dispersion than to measure the relative dispersion, i.e. to measure the dispersion as a ratio or percentage of the average instead of in units of length, weight etc. A standard deviation of one ounce would matter far less in weighing out 7 lb boxes than it would for $\frac{1}{4}$ lb packets of tea.

The commonest measure of relative dispersion is the *Coefficient of Variation*, $v$, obtained by dividing the SD by the mean and expressing it as a percentage. In short,

$$v = 100\frac{\sigma}{\bar{x}} \tag{11.4}$$

The coefficient of variation is not dependent on choice of units. Indeed, it can be used to compare the dispersion of distributions expressed in different units, e.g. the daily output of two factories producing different commodities, one expressed in tons and the other in yards.

Other similar measures are MD/mean or MD/median and $(Q_3 - Q_1)/(Q_3 + Q_1)$. These are rarely, if ever, used in practice.

*Example* 11.5

Brown has a batting average of 32 with a SD of 13. Green has an average of 47 with a SD of 18. Which is the more consistent player?

It is easily shown that $13/32 > 18/47$. Hence Green has the smaller coefficient of variation and may be considered more consistent than Brown.

The coefficient of variation and similar measures cannot be used if the origin is arbitrary. Thus it would be ridiculous to use the coefficient of variation to compare movements of temperature in towns with different mean temperatures. Indeed, for any one place, the result would differ according to the scale used, e.g. Centigrade, Fahrenheit or absolute, because $0°$ is not the same temperature in the three cases. In fact, this coefficient only makes sense if all items are positive magnitudes.

**Skewness**

Even when the average and dispersion are known it is possible to fit many distributions of different shapes to them. One question that

**Fig. 31**   Positively skew distribution showing positions of Mean $(M_1)$, Median $(M_2)$ and Mode $(M_3)$

sometimes arises is: 'Is the distribution symmetrical or lopsided and, if the latter, how lopsided is it?' The answer is generally given in the form of a *coefficient of skewness*, an abstract quantity independent of the unit in which the variable is measured.

Fig. 31 (which is a repetition of Fig. 28) shows a typical skew distribution, with a longer 'tail' on the right of the mode than on the left. In this case the skewness is said to be positive.

When the longer tail is on the left the skewness is negative. When skewness is positive the mean ($M_1$), median ($M_2$) and mode ($M_3$) are in that order, starting from the right, as in Fig. 31.

The more lopsided the distribution, the more will the mean, median and mode differ, so one possible coefficient of skewness is given by

$$j_1 = \frac{M_1 - M_3}{\sigma}, \tag{11.5}$$

or, since $M_1 - M_3 = 3(M_1 - M_2)$ approximately,

$$j_1 = \frac{3(M_1 - M_2)}{\sigma} \tag{11.6}$$

An alternative measure is based on the fact that in a skew distribution the median is not exactly midway between the quartiles. The coefficient is given by

$$j_2 = \frac{Q_1 + Q_3 - 2M_2}{Q_3 - Q_1} \tag{11.7}$$

The latter has the advantage that it must lie between $-1$ and $+1$, whereas $j_1$ has no such convenient limits. There are more theoretically satisfactory ways of measuring skewness, but they have no place in elementary work.

Examples on skewness are often based on J-shaped distributions, which are so obviously skew that there is no point in measuring the skewness. In any case, formulae (11.5) and (11.6) do not apply to such distributions, but principally to the moderately skew 'cocked-hat' type. In the author's opinion, however, measures of skewness are mainly of academic interest.

## Exercises

11.1

Unemployment in Great Britain—July 1974

| Duration in weeks | Number of unemployed (1000s) |
|---|---|
| One or less | 53 |
| Over   1 up to   2 | 41 |
| Over   2 up to   3 | 28 |
| Over   3 up to   4 | 21 |
| Over   4 up to   8 | 57 |
| Over   8 up to 13 | 43 |
| Over 13 up to 26 | 65 |
| Over 26 up to 39 | 37 |
| Over 39 up to 52 | 24 |
|         52 and over | 112 |

Calculate the median and quartile deviation of the above data.

Explain why the median and quartile deviation are preferred to the mean and standard deviation for the above distribution.        (*IoS*)

11.2   The following table indicates the marks obtained by 100 students in an examination:

| Marks | Frequency |
|---|---|
| 55–  64 | 1 |
| 65–  74 | 2 |
| 75–  84 | 9 |
| 85–  94 | 22 |
| 95–104 | 33 |
| 105–114 | 22 |
| 115–124 | 8 |
| 125–134 | 2 |
| 135–144 | 1 |

Using linear interpolation, calculate the median mark obtained. Describe the type of distribution for which the median is the best measure of location, and indicate whether the above distribution is of this type. Also calculate the upper and lower quartiles.        (*IoS*)

11.3   Plot the following figures on a cumulative frequency graph and from this find the value of all the deciles of the distribution.

Distribution of a sample of women by stature

| Stature | Number in sample |
|---|---|
| under 4′ 9″ | 39 |
| 4′ 9″ and 4′ 10″ | 273 |
| 4′ 11″ and 5′ 0″ | 798 |
| 5′ 1″ and 5′ 2″ | 1176 |
| 5′ 3″ and 5′ 4″ | 1206 |
| 5′ 5″ and 5′ 6″ | 584 |
| 5′ 7″ and 5′ 8″ | 217 |
| 5′ 9″ and above | 56 |
| *Total* | 4349 |

(Source: *Women's Measurements and Sizes* (HMSO, 1957)

How will the values of the deciles help a clothing manufacturer decide on a sizing system?                                          (*ICMA*)

11.4   Calculate the mean deviation of the following frequency distribution.

| Weight of fowl (lb) | Number of fowls |
|---|---|
| 0·0– 4·9 | 3 |
| 5·0– 9·9 | 9 |
| 10·0–14·9 | 15 |
| 15·0–19·9 | 30 |
| 20·0–24·9 | 18 |
| 25·0–29·9 | 12 |
| 30·0–34·9 | 9 |

Indicate what your answer conveys about the weight of birds in this distribution.                                                         (*ACA*)

11.5   Number of children of present marriage according to date of marriage.

| | Date of marriage | |
|---|---|---|
| No. of children | 1949 or earlier | 1955–1959 |
| 0 | 3% | 6% |
| 1 | 20% | 11% |
| 2 | 31% | 37% |
| 3 | 19% | 28% |
| 4 | 13% | 11% |
| 5 | 6% | 5% |
| 6* or more | 8% | 2% |
| Total | 100% | 100% |

*Estimated average size of a family of 6 or more children is 6·5.

(Source: *General Household Survey—1971*)

(*a*) For each distribution calculate the mean and standard deviation.

(*b*) Comment, briefly, about the two distributions and your results.

(*IoS*)

11.6   A random sample was taken of 200 candidates who had submitted scripts in a certain question paper at a public examination. The results obtained by the 200 candidates in the examination were as follows:

| Marks obtained out of a possible 100 | Number of candidates |
|---|---|
| Under 20 | 0 |
| 20–29 | 6 |
| 30–39 | 18 |
| 40–49 | 41 |
| 50–59 | 68 |
| 60–69 | 49 |
| 70–79 | 14 |
| 80–89 | 4 |
| 90 and over | 0 |
| Total | 200 |

Calculate:

(i) the mean;

(ii) the standard deviation.

(*ICMA*)

11.7   A sample of 35 values has mean 81·52 and standard deviation 3·74. A second sample of 85 values from the same population has mean 82·31 and SD 3·62. Find the mean and SD of the combined sample of 120 values. (*IoS*)

11.8   The following figures give sets of marks for 10 candidates in two examinations, the English paper being marked out of 20 and the Arithmetic paper out of 50. Using an appropriate measure, find whether there is relatively more variation in the marks for one examination than in those for the other.

| Candidate | English marks | Arithmetic marks |
|-----------|---------------|------------------|
| 1 | 17 | 45 |
| 2 | 8 | 30 |
| 3 | 4 | 9 |
| 4 | 20 | 25 |
| 5 | 11 | 17 |
| 6 | 19 | 33 |
| 7 | 5 | 41 |
| 8 | 16 | 24 |
| 9 | 10 | 37 |
| 10 | 7 | 15 |
| Total | 117 | 276 |

(*IoS*)

# 12 Probability and some special distributions

Probability is of fundamental importance in the theory of statistics, but it can only be treated here in a very elementary manner. The reader who is interested in the subject will find it dealt with more thoroughly in textbooks on algebra or in more advanced books on statistical theory.

Books or chapters on probability deal very largely with problems in coin-tossing, dice-throwing and games of chance, not because it is a trivial or frivolous subject but because the principles of probability are most easily explained by means of such problems. One great advantage they have is that the chances of, say, heads and tails can be assumed known, whereas in most practical problems the probabilities can only be estimated. Thus, a man's expectation of life can be calculated only from current mortality rates, which may not apply to future generations, or from assumed rates in which some arbitrary allowance is made for future improvements.

## Definitions and fundamental laws

In discussing events which can have either of two possible results, it will be convenient to refer to these results as success and failure. The probabilities of success and failure are usually denoted by $p$ and $q$ respectively. (N.B. The terms 'success' and 'failure' are used solely for convenience and do not necessarily mean that $p$ is the probability of a favourable outcome. It could, for example, denote the probability of an item being defective.)

If an event can result in $n$ different outcomes, all equally likely, of which $l$ are successes and $m$ are failures,

$$p = \frac{l}{n} \text{ and } q = \frac{m}{n}$$

Since

$$l + m = n,$$

$$p + q = 1 \tag{12.1}$$

For example, if a playing-card is drawn at random from a full pack, the chance of drawing a spade is $\frac{13}{52}$ ($=\frac{1}{4}$), and the chance of drawing a card other than a spade is $\frac{39}{52}$ ($=\frac{3}{4}$).

More generally, if an event can have one of $n$ possible results and their respective probabilities are $p_1, p_2, \ldots p_n$,

$$p_1 + p_2 + \ldots + p_n = 1 \tag{12.2}$$

It may be taken as self-evident that if $p_1$ is the probability of result $A$, and $p_2$ that of result $B$, where $A$ and $B$ are *mutually exclusive* (i.e. cannot happen together), then the probability of 'either $A$ or $B$' is $p_1 + p_2$. Similarly for three or more results. This is the *addition law* of probability.

Thus, if a number is being chosen at random from the numbers 1 to 100, the chance of picking an even number is $\frac{1}{2}$, that of picking a number ending in 5 is $\frac{1}{10}$, and that of picking either an even number or a number ending in 5 is $\frac{1}{2} + \frac{1}{10} = \frac{3}{5}$. It would, however, be untrue to say that the chance of picking either an even number or a multiple of 5 was $\frac{1}{2} + \frac{1}{5}$, because both events include numbers ending in 0, i.e. they are not mutually exclusive. In plain English, they overlap.

The other basic law is the *multiplication law* of probability, which states that if $p_1, p_2, p_3 \ldots p_n$ are the probabilities of success in $n$ independent events, the probability of success in all together is $p_1 p_2 p_3 \ldots p_n$. Thus, if the probability that a man aged 60 will survive ten years is $\frac{3}{4}$ and that of a woman 50 surviving ten years is $\frac{8}{9}$, the probability that they will both survive ten years is $\frac{3}{4} \times \frac{8}{9} = \frac{2}{3}$.

It is assumed here that the events are independent, i.e. that the result of one does not affect the result of another. Successive tosses of a coin are independent, for however many times it may come down 'heads', the probability of the next result being 'heads' is still $\frac{1}{2}$ (unless it is biased in some way); and if $A$'s neighbour wins a football pool, $A$'s chance of winning it the following week is neither increased nor diminished. But if cards are drawn from a pack and not replaced, the chance of drawing a red card at any particular stage will depend on the colours of the cards already drawn.

### Conditional probability

The multiplication law is easily adapted to dependent events; this is best shown by an example.

*Example* 12.1

A class contains 21 boys and 15 girls, assumed to be equal in attainments. Find the probability that a boy will take first prize and a girl second prize.

The probability that a boy will take first prize is $\frac{21}{36}$. The probability then that, out of the other 35 children, one of the 15 girls will take second prize, is $\frac{15}{35}$. The probability of the double event is therefore

$$\frac{21}{36} \times \frac{15}{35} = \frac{1}{4}$$

An alternative argument is as follows: a boy can be chosen for first prize in 21 ways, and a girl for second prize in 15 ways, giving a total of $21 \times 15$ ways. Disregarding sex, the two prizes can be awarded in $36 \times 35$ ways. The required probability is therefore

$$\frac{21 \times 15}{36 \times 35} = \frac{1}{4} \text{ as before.}$$

To solve the great majority of problems in probability, the reader needs only the above basic theorems, a logical mind, and plenty of practice in working out examples.

### The probability of *r* successes in *n* independent and similar events

The present section is of great importance in the theory of sampling. Suppose an event is repeated and that the probability $p$ of success is the same each time. The problem is to find the respective probabilities of $0, 1, 2, \ldots n$ successes in $n$ trials.

To take a concrete case, suppose a six-sided die is thrown repeatedly, and that a six counts as a success, any other value being a failure, so that $p = \frac{1}{6}$ and $q = \frac{5}{6}$.

Consider first the case of two throws—or two dice thrown together, which comes to the same thing. By the multiplication law, the probability of two successes is $(\frac{1}{6})^2$ and that of two failures is $(\frac{5}{6})^2$. The probability that the first throw is a success and the second a failure is $\frac{1}{6} \times \frac{5}{6}$. The probability that the first throw is a failure and the second a success is also $\frac{1}{6} \times \frac{5}{6}$. Hence the probability of one success and one failure, regardless of order, is $2 \times \frac{1}{6} \times \frac{5}{6}$.

It is easily verified that

$$\left(\frac{1}{6}\right)^2 + \left(\frac{5}{6}\right)^2 + 2 \left(\frac{1}{6}\right)\left(\frac{5}{6}\right) = 1$$

On similar reasoning, the chances of 0, 1, 2, 3 successes in three throws are respectively

$$\left(\frac{5}{6}\right)^3, \quad 3 \left(\frac{5}{6}\right)^2 \left(\frac{1}{6}\right), \quad 3 \left(\frac{5}{6}\right)\left(\frac{1}{6}\right)^2, \quad \left(\frac{1}{6}\right)^3$$

In general, if there are $n$ trials, and $p$ and $q$ are the chances of success and failure in each trial, the probability of $r$ successes and $(n-r)$ failures *in a specified order* is $p^r q^{n-r}$. But since $r$ successes can occur in $^nC_r$ ways, the probability of $r$ successes regardless of their order is $^nC_r p^r q^{n-r}$, where

$$^nC_r = \frac{n!}{r!(n-r)!} = \frac{n(n-1)\ldots(n-r+1)}{1.2.\ldots r}$$

Those familiar with the Binomial Theorem will recognize the probabilities of 0, 1, 2, ... $n$ successes as the respective terms in the expansion of $(q + p)^n$, viz.

$$q^n, \; npq^{n-1}, \; \ldots \; ^nC_r p^r q^{n-r}, \; \ldots p^n \tag{12.3}$$

*Example* 12.2

Find the probability that out of 10 coins tossed, exactly 3 will be 'heads'.

Since $p = q = \frac{1}{2}$, the required probability is

$$^{10}C_3 (\tfrac{1}{2})^3 (\tfrac{1}{2})^7 = \frac{10.9.8}{1.2.3} (\tfrac{1}{2})^{10}$$

$$= \frac{15}{128}$$

## Probability distributions

The expressions given by formula (12.3) provide an example of a *probability distribution*, which is a series of probabilities corresponding to all possible values of a variable, in this case the number of successes in $n$ events. Probability and frequency distributions are closely connected, for if a large number of experiments are performed, the relative frequencies of the various possible results will approximate to the probabilities. A probability distribution may therefore be regarded as a frequency distribution standardized so that the total frequency is unity, and it can be represented by a histogram, polygon or curve like any other frequency distribution.

**Fig. 32**  Probability distribution of total scores made by throwing three dice

Fig. 32 shows, in the form of a frequency polygon, the probability distribution of scores of 3, 4, ... 18 with three dice thrown simultaneously. The actual probabilities are

$$\frac{1}{216}(1, 3, 6, 10, 15, 21, 25, 27, 27, 25, 21, 15, 10, 6, 3, 1)$$

The fraction outside the bracket means that all numbers inside the bracket should be divided by 216. The calculation of these probabilities is rather laborious.

When the variable is continuous, it is possible to compile and plot a probability distribution, but the interpretation is not quite so simple. It is, however, possible to estimate the probability that the height or weight of a man chosen at random lies between certain limits, or that a machine will break down within a stated time.

**Fig. 33**  Probability distribution of times taken for a machine to break down

Fig. 33 shows a possible distribution for times run by a machine before it breaks down. Since the total of all probabilities must be unity, the curve is scaled so that the area between it and the axes is also unity. The area *PNMQ* then represents the probability that the machine will break down between $1\frac{1}{2}$ and 2 hours, the times represented by *N* and *M*. The area to the left of *PN* gives the probability of its breaking down within $1\frac{1}{2}$ hours, and the 'tail' to the right of *QM* that of the machine running at least 2 hours without a breakdown.

Unfortunately, these areas cannot be read off as vertical heights can, and before a continuous probability distribution can be of much use it must be tabulated. Several such distributions, like the so-called Normal Distribution (page 131), have been extensively tabulated, as also have some discrete distributions, so that the probability of the variable taking certain values or lying within or outside a given range can be found with reasonable accuracy.

## Expectation

The mean of a probability distribution is often called the *expected value*, or *expectation*, of the variable *x*, and is denoted by $E(x)$. If *x* can take values $x_1, x_2, \ldots x_n$ with respective probabilities $p_1, p_2, \ldots p_n$, then:

$$E(x) = p_1 x_1 + p_2 x_2 + \ldots + p_n x_n \tag{12.4}$$

It is easily seen that $E(x)$ will be the average value of $x$ in a very large number of observations.

'Expected value' is rather a misnomer, since it may be a value the variable cannot possibly take. For example, if five coins are tossed, the 'expected number' of heads is $2\frac{1}{2}$. The term 'expectation' is preferable for this reason.

## Example 12.3

A boy answers four questions 'yes' or 'no' and guesses each time, not knowing the answers. He is to receive 5p for 2 correct answers, 15p for 3, and 30p for 4. Find his expectation.

From formula (12.3), since $p = q = \frac{1}{2}$,

$$\text{probability of 2 correct answers} = {}^4C_2(\tfrac{1}{2})^4 = \tfrac{3}{8}$$
$$\text{,,} \quad \text{,, 3} \quad \text{,,} \quad \text{,,} = {}^4C_3(\tfrac{1}{2})^4 = \tfrac{1}{4}$$
$$\text{,,} \quad \text{,, 4} \quad \text{,,} \quad \text{,,} = {}^4C_4(\tfrac{1}{2})^4 = \tfrac{1}{16}$$

$\therefore$ expectation in pence

$$= \tfrac{3}{8} \times 5 + \tfrac{1}{4} \times 15 + \tfrac{1}{16} \times 30$$
$$= \tfrac{15}{8} + \tfrac{15}{4} + \tfrac{15}{8}$$
$$= 7\tfrac{1}{2}\text{p}$$

## The binomial distribution

One of the commonest probability distributions is the one already given in formula (12.3) by the binomial expansion of $(q + p)^n$ and called the *Binomial Distribution*. It is really a family of distributions, the shape depending on the values of $n$ and $p$.

When the chances of success and failure are equal, i.e. when $p = q = \frac{1}{2}$, the probability of $x$ successes is ${}^nC_x(\frac{1}{2})^n$, and equal to that of $x$ failures, or $(n-x)$ successes. Obviously the distribution is symmetrical about $x = \frac{1}{2}n$.

When $p$ and $q$ are unequal, the distribution is skew, but if $n$ is fairly large and neither $p$ nor $q$ is very small, the skewness is only slight and the tails can almost be ignored because the extreme frequencies are so small. If $p$ is smaller than $q$, the longer tail is on the right.

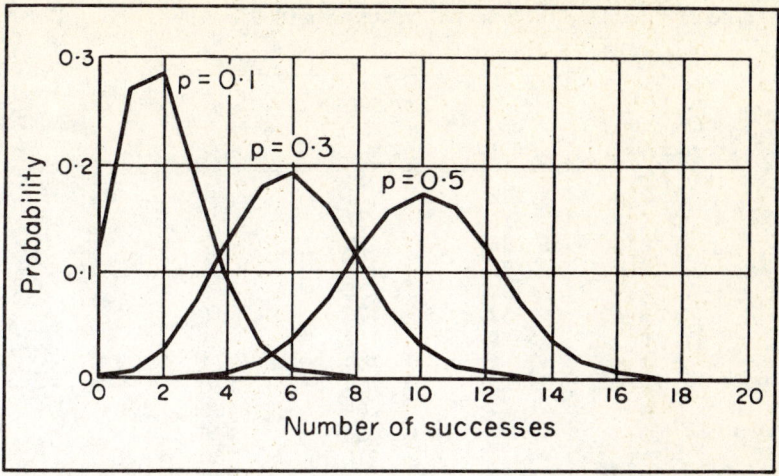

**Fig. 34** Probability distribution of 0, 1, 2, . . . 20 successes in twenty trials when $p = 0.1$, $0.3$ or $0.5$

Fig. 34 shows distributions corresponding to the expansion of $(q + p)^{20}$ when $p = 0.1, 0.3$ and $0.5$. The first $(p = 0.1)$ is markedly skew, the second almost symmetrical, and the last perfectly symmetrical. The actual probabilities, multiplied by 1000 to avoid decimals, are given in Table 21.

The mean of the binomial distribution is $np$, as is evident from the very definition of $p$. It will be shown later (page 146) that the standard deviation is given by the formula

$$\sigma^2 = npq \tag{12.5}$$

This formula is extremely important, as the following chapter will show.

## The Normal distribution

We come now to a continuous distribution of supreme importance in the theory of statistics, called the *Normal Distribution,* or sometimes the *Gaussian Distribution.* The curve representing it is called the *Normal Curve* (see Fig. 35), and its equation, for the benefit of the mathematically minded students, is, taking the mean as zero,

$$y = \frac{1}{\sigma\sqrt{(2\pi)}} e^{\frac{-x^2}{2\sigma^2}} \tag{12.6}$$

**Table 21**   Terms of the binomial expansion of $1000 (q + p)^{20}$

(Figures to nearest unit)

| Number of successes | Frequency | | |
|---|---|---|---|
| | $p = 0 \cdot 1$ | $p = 0 \cdot 3$ | $p = 0 \cdot 5$ |
| 0 | 122 | 1 | — |
| 1 | 270 | 7 | — |
| 2 | 285 | 28 | — |
| 3 | 190 | 72 | 1 |
| 4 | 90 | 130 | 5 |
| 5 | 32 | 179 | 15 |
| 6 | 9 | 192 | 37 |
| 7 | 2 | 164 | 74 |
| 8 | — | 114 | 120 |
| 9 | — | 65 | 160 |
| 10 | — | 31 | 176 |
| 11 | — | 12 | 160 |
| 12 | — | 4 | 120 |
| 13 | — | 1 | 74 |
| 14 | — | — | 37 |
| 15 | — | — | 15 |
| 16 | — | — | 5 |
| 17 | — | — | 1 |
| 18 | — | — | — |
| 19 | — | — | — |
| 20 | — | — | — |

The shape is always the same: it is symmetrical about the mean and stretches away to infinity on either side, although in practice it is indistinguishable from the base line at distances greater than $3\sigma$ from the mean. The two points of inflexion occur exactly one standard

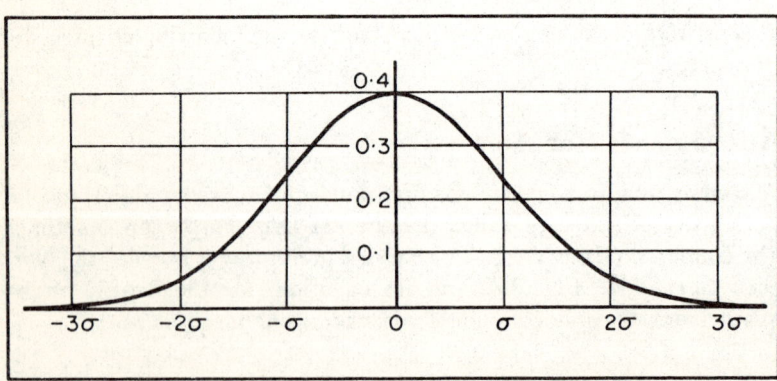

**Fig. 35**   Normal curve

deviation from the mean on either side. The QD is about $0.6745\sigma$, and the MD about $0.7979\sigma$.

The Normal Curve was studied by Gauss in connection with the theory of errors of observation, and is sometimes called the Curve of Errors. It is a kind of ideal curve which very nearly fits many of the distributions met in statistics, such as distributions of men's heights, examination marks and errors of observation. The student should, however, beware of statements sometimes made that the Normal Curve is found in 'natural data'. Very often a Normal Curve can be fitted to data, and unless they are very numerous it may not be possible to prove that the distribution is not Normal, but perfect Normality rarely occurs in practice.

As equation (12.6) shows, the Normal Distribution with zero mean has the great advantage of containing only one arbitrary constant $\sigma$ (the SD), which means that it is easy to tabulate. This can be done in various ways. Table 22 shows the proportion $P$ of the area of the curve

**Table 22** Proportion of the area of the Normal Curve to the right of the ordinate at deviation $x$ from the mean

| $x/\sigma$ $(=\mu)$ | $P$ | $x/\sigma$ $(=\mu)$ | $P$ | $x/\sigma$ $(=\mu)$ | $P$ |
|---|---|---|---|---|---|
| 0 | 0·5000 | 1·0 | 0·1587 | 2·0 | 0·0228 |
| 0·1 | 0·4602 | 1·1 | 0·1357 | 2·1 | 0·0179 |
| 0·2 | 0·4207 | 1·2 | 0·1151 | 2·2 | 0·0139 |
| 0·3 | 0·3821 | 1·3 | 0·0968 | 2·3 | 0·0107 |
| 0·4 | 0·3446 | 1·4 | 0·0808 | 2·4 | 0·0082 |
| 0·5 | 0·3085 | 1·5 | 0·0668 | 2·5 | 0·0062 |
| 0·6 | 0·2743 | 1·6 | 0·0548 | 2·6 | 0·0047 |
| 0·7 | 0·2420 | 1·7 | 0·0446 | 2·7 | 0·0035 |
| 0·8 | 0·2119 | 1·8 | 0·0359 | 2·8 | 0·0026 |
| 0·9 | 0·1841 | 1·9 | 0·0287 | 2·9 | 0·0019 |
| | | | | 3·0 | 0·00135 |

that lies to the right of the ordinate at various distances from the mean. This area represents the probability that an item taken at random from the distribution exceeds the mean by that amount ($x$) or more.

The mean of the Normal Distribution (or of any other distribution, for that matter) is commonly denoted by $\mu$ (Greek letter 'mu'). The factor $x/\sigma$ is called a *Normal deviate* and is often denoted by $u$. From the symmetry of the curve it is clear that the area to the right of $\mu + u\sigma$ equals the area to the left of $\mu - u\sigma$. It is important to be able to use the tables of the Normal Distribution easily and correctly. Examples 12.4 and 12.5 illustrate how they can be used.

*Example* 12.4

Weights of mass-produced articles are Normally distributed with mean 500 g and SD 15 g. Find what proportions of the whole output weigh (i) over 530 g, (ii) less than 488 g.

(i) 530 g is $2\sigma$ more than the mean, and from Table 22 the 'tail' (i.e. the value of $P$) corresponding to $u = 2$ is $0\cdot0228$, or $2\cdot28$ per cent.

(ii) 488 g is 12 g or $0\cdot8\sigma$ less than the mean, and the tail to the left of $u = -0\cdot8$ is equal to that to the right of $u = +0\cdot8$, viz. $0\cdot2119$, or $21\cdot19$ per cent.

*Example* 12.5

A process produces rods varying slightly in length with SD $0\cdot3$ mm. The machine can be set to adjust the average length. What should this be if not more than 1 per cent of all rods are to be shorter than 15 cm? State any assumptions you make.

(N.B. The last sentence is a favourite with examiners. Nine times out of ten, one such assumption, and often the only one required, is that the distribution is Normal.)

Figure 36 will probably make the problem clearer to the student. The tail to the left of 15 cm must be 1 per cent or less. Interpolating between $\mu = 2\cdot3$ and $\mu = 2\cdot4$ we find that $P$ is just under $0\cdot01$ when $\mu = 2\cdot33$. Hence, $\mu$ should exceed 15 cm by $2\cdot33\sigma$, i.e. $\mu = (15 + 2\cdot33 \times 0\cdot03)\,\text{cm} = 15\cdot07\,\text{cm}$.

**Fig. 36**

A useful device for checking whether a distribution is Normal, or nearly so, is *Normal probability paper*. This is graph paper on which one axis represents the variable and the other the percentage cumulative frequency (i.e. the left-hand tail), the latter being so graduated that the cumulative Normal curve is transformed into a straight line. For some obscure reason, this paper is often printed with the variable plotted vertically and the cumulative frequency horizontally, which is confusing.

Figure 37 shows the data of Table 20 plotted on probability paper. The points fit a straight line very well, showing that the distribution is nearly Normal.

**Fig. 37**  Data of Table 20 plotted on Normal probability paper

The first and last points present a little difficulty, since 0 per cent and 100 per cent exactly cannot be shown, but in any case the extreme frequencies are very small and subject to accidents of sampling, so they can almost be disregarded.

If a line is fitted by eye, the 50, 25 and 75 per cent points provide good estimates of the median (in this case also the mean) and quartiles.

The reader may have noticed a resemblance between the Normal Curve and at least some of the frequency polygons shown in Figs 32 and 34. Many distributions do, in fact, approximate to the Normal when the number of observations is large. In particular, the binomial distribution can be closely fitted by a Normal distribution with the same mean and SD when $n$ is fairly large (say, at least 25) and neither $p$ nor $q$ is very small. This is useful if we wish to calculate the probability of getting at least (or at most) a certain number of successes.

*Example* 12.6

A random sample of 25 articles is taken from a very large batch in which 20 per cent of all articles are defective. Find the probability that the sample will contain 8 or more defectives.

Here we have a binomial distribution with

$$n = 25, p = 0 \cdot 2, q = 0 \cdot 8,$$
$$\mu = np = 5, \sigma = \sqrt{npq} = 2$$

It would be laborious to calculate and add the first 8 terms (or the last 18) of this distribution, so we approximate it by a Normal distribution with $\mu = 5$, $\sigma = 2$. But be careful! A histogram of the binomial distribution would represent 8 defectives by a rectangle on the interval 7·5 to 8·5, so the probability of 8 or more is estimated from the tail of the Normal Curve to the right of 7·5, not 8. This is called a *correction for continuity*.

Then $\mu = (7 \cdot 5 - 5) \div 2 = 1 \cdot 25$, and by interpolation in Table 22 (or from more detailed tables),

$$P = 0 \cdot 106$$

The actual probability, calculated from the binomial distribution, is 0·109.

The proof of the approximation is beyond the scope of this book.

## The Poisson distribution

Another limiting form of the binomial distribution is the *Poisson Distribution*. This is the form to which the binomial distribution approximates when $n$ is large, $p$ very small, and $np$ of the order of unity, say less than 10. It can be shown (the proof will not be given here) that successive terms of the binomial distribution approximate to

$$e^{-m}, me^{-m}, \tfrac{1}{2}m^2e^{-m}, \ldots \frac{m^x}{x!} e^{-m}, \text{ etc.,}$$

where $m = np$ and $e$ denotes the base of the system of natural logarithms ($=2\cdot71828\ldots$). This is really a family of distributions, and the shape depends very much on the value of $m$.

Since the mean and standard deviation of the binomial distribution are $np$ and $\sqrt{(npq)}$ respectively, and $q$ is now nearly equal to 1, it follows that the mean and standard deviation of the Poisson Distribution are respectively $m$ and $\sqrt{m}$.

This distribution, like the Normal Distribution, has the advantage that it involves only one parameter, and can therefore be easily tabulated. It is useful when the expected number of successes or failures in a large number of events is small. For example, if the average number of defective articles in fairly large batches of a given size is 2, the probability of finding no defectives in a single batch is $e^{-2}$, the probability of finding exactly one is $2e^{-2}$, and so on. Each term after the first can be derived from the previous term by multiplying successively by 2, $\frac{2}{2}$, $\frac{2}{3}$, $\frac{2}{4}$, etc., as the reader can verify from the following table. For most values of $m$, however, the construction is more troublesome.

| Number of defectives | Probability |
|---|---|
| 0 | 0·1353 |
| 1 | 0·2707 |
| 2 | 0·2707 |
| 3 | 0·1805 |
| 4 | 0·0902 |
| 5 | 0·0361 |
| 6 | 0·0120 |
| 7 | 0·0034 |
| 8 | 0·0009 |
| 9 | 0·0002 |

The Poisson Distribution may arise in any situation where an event is liable to occur at random at irregular intervals of time or space, and the probability of occurrence in a small interval of given length or size is constant. Examples are:

(a) The number of accidents occurring in a given factory during a day or week.

(b) The number of cars passing a certain point in one minute.

(c) The number of telephone calls received on one line during a given short period.

(d) The number of organisms on a square of a haemocytometer.

In all these examples it is assumed that conditions, such as weather, which have any bearing on the event being observed are constant over the period or interval in question.

*Example* 12.7

A machine breaks down, on the average, once a week, at random times. What is the probability that it will break down at least three times in the next fortnight?

The expected number of breakdowns in a fortnight is 2, so from the above table the probability of 3 or more is

$$1 - (0 \cdot 1353 + 0 \cdot 2707 + 0 \cdot 2707)$$
$$= 0 \cdot 3233$$

Notice that it is assumed that breakdowns occur independently of one another, so that the probability of a breakdown in the next minute is unaffected by the length of time since the last breakdown.

**Exercises**

12.1   (*a*) A pack of cards is cut at random, one card exposed, the suit of the card noted and the whole pack re-made. If this procedure is performed four times, what is the probability that none of the exposed cards is a spade?

(*b*) In an office, four letters and the four corresponding envelopes are typed. If the letters are put into the envelopes at random, what is the chance that none of the letters is put into the correct envelope?

(*IoS*)

12.2   The probability that machine A will be performing a useful function in 5 years' time is $\frac{1}{4}$ while the probability that machine B will still be operating usefully at the end of the same period is $\frac{1}{3}$.

Find the probability that in five years' time:

(*a*) both machines will be performing a useful function;
(*b*) neither will be operating;
(*c*) only machine B will be operating;
(*d*) at least one of the machines will be operating.       (*ICMA*)

12.3   The mean daily output of operatives in a spinning factory is Normally distributed with a mean of 540 yards and a standard deviation of 17 yards. The management are introducing more modern machines but will only train workers whose mean daily output is already 525 yards or more to use them. What percentage of the factory force can expect re-training? (*IoS*)

12.4 The mean weight of a consignment of 500 sacks of sugar is 151 lb and the standard deviation 15 lb. Assuming that the weights are Normally distributed, find how many sacks weigh:

(*a*) between 120 and 155 lb,
(*b*) more than 185 lb, and
(*c*) less than 128 lb. (*ACA*)

12.5 (*a*) Fit a Normal Curve to the following data:

| Weight (nearest 1 lb) | Number of men |
|---|---|
| Under 100 | 6 |
| 100–119 | 43 |
| 120–139 | 93 |
| 140–159 | 191 |
| 160–179 | 263 |
| 180–199 | 212 |
| 200–219 | 144 |
| 220–239 | 40 |
| Over 239 | 8 |
| Total | 1000 |

(*b*) Estimate the weight which is exceeded by 10 per cent of the men. (*IoS*)

12.6 It is specified that the bursting strength of a particular type of paper should be at least 25 pounds per square inch (psi). Current production is such that the mean bursting strength is at 26 psi, with a standard deviation of 0·75 psi.

(*a*) Assuming that bursting strength is distributed Normally, what proportion of production is outside specification?
(*b*) What percentage increase in bursting strength should be required to ensure that only 1 per cent of production is outside specification? (The standard deviation may be assumed to be unchanged.) (*IoS*)

12.7  (*a*) The mean inside diameter of a sample of 400 washers produced by a machine is 8·92 millimetres and the standard deviation is 0·12 millimetres. Washers with inside diameters within the range of 8·74 to 9·10 millimetres are acceptable.

Calculate the percentage of defective washers produced by the machine, assuming the diameters are Normally distributed.

(*b*) Five per cent of the units produced in a manufacturing process turn out to be defective.

Find the probability that in a sample of ten units chosen at random exactly two will be defective using:

  (i) the binomial distribution,
  (ii) the Poisson approximation to the binomial distribution.

(*ICMA*)

12.8  It is expected that 10 per cent of the production from a continuous process will be defective and scrapped.

Determine the probability that in a sample of ten units chosen at random:

  (*a*) exactly two will be defective; and
  (*b*) at the most two will be defective;

using in each case both:

  (i) the binomial distribution; and
  (ii) the Poisson approximation to the binomial distribution.

(*ICMA*)

12.9  (*a*) Under what conditions can the binomial distribution be approximated by the Poisson distribution, and in what circumstances is such an approximation useful?

(*b*) If the proportion of defectives in a bulk is 4 per cent, find the probability of not more than one defective in a sample of 10.

(*c*) A shopkeeper's sales of a certain article amount to 4 per month on the average, the sales occurring at random and independently. He replenishes his stock once a month. To what number should he make up his stock in order to reduce the chance of running out of stock to less than 1/100?  (*IoS*)

12.10  A rolling machine in a paper mill produces on average one flaw every 500 feet of paper. Assuming that the number of flaws in a given length of paper has a Poisson distribution, find the probability that

(*a*) a 2000 ft roll has no flaws;

(*b*) a 500 ft roll has at most three flaws;

(*c*) out of three 1000 ft rolls, one has one flaw and the others have none. *(IoS)*

# 13 Sampling

In Chapter 4 the technique of sampling was described in general terms, with special reference to sample inquiries of the social survey type. It was stated that the larger the sample and the more carefully it was chosen, the more reliable it would be as a cross-section of the population, but the numerical expression of the reliability of samples was deferred to this chapter.

The theory of sampling is generally divided into the sampling of attributes and the sampling of variables. In the sampling of attributes, the first problem is to estimate the proportions of items falling into categories that cannot be expressed numerically, like brown-eyed, blue-eyed etc., male and female, or alive and dead. In the sampling of variables the problem is to estimate the mean, and perhaps the dispersion, of some variable like height or weight or the yield of a product. In both cases, subsequent problems are to assess the precision of these estimates and to judge the significance of differences between one estimate and another, or between an estimate and some expected value.

Throughout this chapter it will be assumed that all samples are random samples unless otherwise stated.

## Standard deviation of a sum, difference, or mean

Suppose there are two variables $x$ and $y$, each with its own distribution and with means $\mu_x$ and $\mu_y$ and standard deviations $\sigma_x$ and $\sigma_y$ respectively. Suppose further that an item is taken at random from each distribution and that the two items are added together, giving $z = x + y$. Then $z$ will form another distribution with mean $\mu_x = \mu_y \, \mu_z$. The problem now is to find $\sigma_z$, the standard deviation of $z$.

To simplify the calculations, let $x$, $y$ and $z$ be measured from their respective means, and suppose $x$ and $y$ can each take a limited number of values $x_1$, $x_2$, ... $y_1$, $y_2$ etc., with probabilities $p_1$, $p_2$, ... $q_1$, $q_2$ etc., respectively. (Students familiar with integral calculus can generalize this section to the case where the distributions of $x$ and $y$ are continuous. The non-mathematician may skip the proof and continue from formula (13.2).)

Then if $z_{rs} = x_r + y_s$, the probability associated with $z_{rs}$ is $p_r q_s$. Since $x$, $y$ and $z$ are measured from their means,

$$\Sigma_r p_r x_r = \Sigma_s q_s y_s = \Sigma_r \Sigma_s p_r q_s z_{rs} = 0,$$
$$\sigma_x^2 = \Sigma_r p_r x_r^2 \text{ and } \sigma_y^2 = \Sigma_s q_s y_s^2 \tag{13.1}$$

Then

$$\begin{aligned}
\sigma_z^2 &= \Sigma_r \Sigma_s p_r q_s z_{rs}^2 \\
&= \Sigma_r \Sigma_s p_r q_s (x_r + y)^2 \\
&= \Sigma_r \Sigma_s p_r q_s x_r^2 + 2\Sigma_r \Sigma_s p_r q_s x_r y_s + \Sigma_r \Sigma_s p_r q_s y_s^2 \\
&= (\Sigma_r p_r x_r^2)(\Sigma_s q_s) + 2(\Sigma_r p_r x_r)(\Sigma_s q_s y_s) + (\Sigma_r p_r)(\Sigma_s q_s y_s^2) \\
&= (\sigma_x^2 \times 1) + (2 \times 0 \times 0) + (\sigma_y^2 \times 1) \qquad \text{(from (13.1))} \\
&= \sigma_x^2 + \sigma_y^2 \tag{13.2}
\end{aligned}$$

In other words, the variance of $z(=x + y)$ is equal to the sum of the variances of $x$ and $y$, provided that $x$ and $y$ are independent.

This extremely important formula can be extended to any number of distributions, irrespective of their form or the numbers of items they contain. Furthermore, it applies equally to differences, as can be seen by changing the sign of $y$ in formula (13.1). Thus if

$$d = x - y,$$
$$\sigma_d^2 = \sigma_x^2 + \sigma_y^2 \text{ as before,} \tag{13.2a}$$

and if

$$X = x_1 \pm x_2 \pm \ldots \pm x_n,$$
$$\sigma_X^2 = \sigma_1^2 + \sigma_2^2 + \ldots + \sigma_n^2 \tag{13.3}$$

This result can be generalized to any linear function of the variables. If

$$T = a_1 x_1 + a_2 x_2 + \ldots + a_n x_n$$

where $a_1$, $a_2$, ... $a_n$ are constants and $x_1$, ... $x_n$ are all independent of one another, then

$$\sigma_T^2 = a_1^2 \sigma_1^2 + a_2^2 \sigma_2^2 + \ldots + a_T^2 \sigma_n^2 \tag{13.4}$$

Two very important results can be obtained from equation (13.3) when $x_1, x_2, \ldots x_n$ are all taken *from the same distribution*. Since the standard deviation is now the same for each item ($=\sigma$, say),

$$\sigma_X^2 = n\sigma^2$$

or

$$\sigma_X = \sigma\sqrt{n} \tag{13.5}$$

Now $\bar{x}$ is the mean of the $n$ items selected, $\bar{x} = X/n$, and $\sigma_{\bar{x}} = \sigma_X/n$. Hence

$$\sigma_{\bar{x}} = \sigma\frac{\sqrt{n}}{n}$$

$$= \frac{\sigma}{\sqrt{n}} \tag{13.6}$$

To see what this means, suppose the heights of a large body of men, such as an army regiment, are distributed with a mean of 69·2 in. and a standard deviation of 2·7 in. Then if a large number of samples of 25 men are taken, the average heights of these samples will themselves form a distribution with a mean of 69·2 in. (the population mean) and a standard deviation not of 2·7 in. but of

$$\frac{2\cdot7}{\sqrt{25}} = \frac{2\cdot7}{5} = 0\cdot54 \text{ in.}$$

If samples of 100 men were taken, the standard deviation would be reduced to 0·27 in. Clearly formula (13.6) is a useful guide to the precision of an estimated average.

## Sampling distributions

Other quantities besides sums, differences and means can be estimated from samples. Sometimes variability is more important than the average, and the sample is then taken to obtain an estimate of the standard deviation; or some measure of skewness may be required to test whether the distribution is nearly Normal. Any such quantity estimated from a sample is called a *statistic*.

For a given size of sample any statistic will have a *sampling distribution* formed by all possible values of the statistic and depending on the number of items in each sample and the form of the parent distribution. For example, if the parent distribution is Normal, it can be shown that the distribution of the mean of $n$ items is also Normal, but the standard deviation depends on the value of $n$ (see Fig. 38). It can also be shown, but only in a more advanced book than this, that for most types of distribution the mean of $n$ items becomes more and more Normally distributed as $n$ increases. So does the total of $n$ items (cf. Fig. 32).

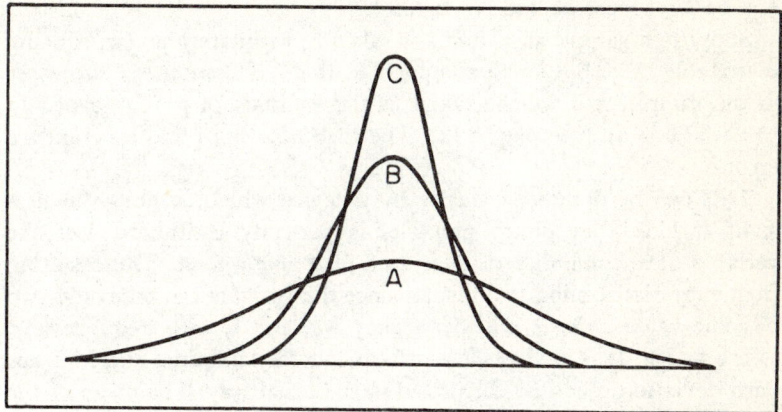

**Fig. 38** Normal distribution ($A$), with distributions of sample means of four items ($B$), and of nine items ($C$)

Note that the parent distribution need not consist of items actually observed. It may be a probability distribution or a purely hypothetical distribution of values that might be observed if observations were continued indefinitely. For example, in an experiment to determine a physical constant, a temperature or the yield of a process under given conditions, the various results obtained by different observers, by different methods and with different apparatus may be regarded as a sample of an infinite population of possible estimates. Such a hypothetical population is often called a *universe*.

The standard deviation of a sampling distribution is generally called the *Standard Error*. Thus in the example given in page 144, the standard error of the mean height of 25 men is 0·54 in. The word 'error' is used in a technical sense to denote the difference between an estimate and the true value, which in this case is the average height of the population, and the standard error is a measure of the uncertainty of that estimate.

### Standard error of a proportion

It is now possible to find the standard error of an estimated proportion, e.g. the proportion of births that result in twins. Consider a large population of $N$ items, divisible into two categories only, which it is convenient to call successes and failures. (The reader can decide for himself into which category he would put twins.) Then if $p$ and $q$ are the (unknown) proportions of successes and failures respectively, there will be $Np$ successes and $Nq$ failures.

Suppose a sample of $n$ items is taken to estimate $p$ and $q$, $n$ being reasonably large but small compared with $N$. If there are $x$ successes in the sample, $x/n$ will be taken as the estimate of $p$. The expected value of $x$ is $np$ (see page 131). The task now is to find its standard error.

This can be done very neatly by a device which, although at first sight it looks like sharp practice, is perfectly legitimate. Let the variable be the number of successes *in a single item*. The resulting frequency distribution is unusual, since the variable can take only two possible values, viz. 0, with frequency $Nq$, and 1, with frequency $Np$ (see Fig. 39). It is nevertheless a frequency distribution, and its standard deviation $\sigma$ can be calculated in the usual way. The mean of the distribution is clearly $p$.

Then

$$N\sigma^2 = Nq(0 - p)^2 + Np(1 - p)^2$$
$$= Nqp^2 + Npq^2$$
$$= Npq(p + q)$$
$$= Npq$$
$$\therefore \sigma^2 = pq$$

But the sample is a sample of $n$ items from this distribution, so from formula (13.5) the standard error of $x$ is given by

$$\sigma_x{}^2 = npq$$
$$\sigma_x = \sqrt{npq} \tag{13.7}$$

Since the observed proportion is $x/n$, its standard error is

$$\frac{1}{n}\sqrt{npq} = \sqrt{\frac{pq}{n}} \tag{13.8}$$

If $n$ is an appreciable fraction of a finite population of size $N$, this formula must be modified to

$$\sqrt{\frac{pq}{n}\frac{N-n}{N-1}} \tag{13.8a}$$

**Fig. 39**  Distribution of successes and failures
in a population of size $N$

The factor $\sqrt{(N-n)/(N-1)}$ is called the *finite population correction factor*. It applies also to the standard error of a mean (13.6).

## Standard errors of median and standard deviation

The standard error of the mean of a sample is very simple and does not depend on the form of the parent distribution. The standard error of the median is more complicated, but for a Normal parent distribution it is approximately $1 \cdot 25\sigma/\sqrt{n}$, compared with $\sigma/\sqrt{n}$ for the mean. In other words, the mean gives a more reliable estimate of the central value than the median, provided that the distribution is Normal, or approximately so.

Sometimes it is necessary to compare standard deviations, e.g. when an attempt is made to reduce the variability of a process. The standard error of a standard deviation is approximately $\sigma/\sqrt{2n}$, provided again that the population is Normal.

The proofs of these two formulae will not be given here. Both formulae may be considerably modified when the parent distribution is far from Normal.

## Confidence limits

It is generally possible, using the standard error of a statistic such as a mean or proportion, to determine a *confidence interval* which we are fairly sure contains the true value of which that statistic is an estimate.

The end-values of that interval are called *confidence limits*. Their actual values will depend on the desired *confidence level*, i.e. our degree of belief that the true value does fall in that interval. Thus, 95 per cent confidence limits are calculated in such a way that we are 95 per cent sure that the true value is between those limits. Putting it another way, if we say that the true value lies between the 95 per cent confidence limits we shall be right in 95 per cent of such statements.

It is sometimes said loosely that there is a 95 per cent probability that the true value lies between the limits, but this is not really correct. The probability is either 1 (if it does) or 0 (if it does not). It is best to restrict the term 'probability' as applying to future events.

As already remarked, the mean of a large sample of observations is almost Normally distributed for most types of parent distribution; so also is the observed proportion of successes, provided neither $np$ nor $nq$ is very small, say less than 10. Consequently the tables of the Normal Distribution may be used with considerable accuracy to calculate the required limits. The following two examples will illustrate the method.

*Example* 13.1

It is desired to estimate the proportion of people in a certain town who have been vaccinated. Out of a random sample of 300, only 75 had been vaccinated. Find the 95 per cent confidence limits for the population.

Substituting $n = 300$, $p = 75/300 = 0.25$, $q = 0.75$ in formula (13.8),

$$\sigma_p = \text{SE of } p = \sqrt{\frac{0.25 \times 0.75}{300}} = 0.025.$$

From the tables of the Normal Distribution, 95 per cent of the observations lie within the range (Mean $\pm 1.96\sigma$). In other words, in 95 per cent of similar occasions, the true value will be within $1.96\sigma$ of our estimate. Hence with 95 per cent confidence, the proportion of the population that has been vaccinated lies between the limits

$$0.25 \pm 1.96 \times 0.025 = 0.25 \pm 0.049$$

i.e. between $0.201$ and $0.299$, or between $20.1$ per cent and $29.9$ per cent.

One difficulty that was glossed over in this example is that since it was the *estimated* values of $p$ and $q$ that were substituted in the formula, $\sigma_p$ itself may not be correct. It is, however, the best estimate that can be found from the sample, and if the true value of $p$ were 30 per cent, $\sigma_p$ would only be increased to $0.0265$. For greater precision a larger sample would of course be necessary.

If it were desired, following this preliminary survey, to estimate with 95 per cent confidence the percentage vaccinated correct to, say, 2 per cent, the size of sample required could be calculated as follows:

$1.96\sigma_p$ must be made equal to $0.02$

$$\therefore (1.96)^2 \frac{pq}{n} = (0.02)^2$$

Putting $p = 0.25$, $q = 0.75$ (the best estimates yet available)

$$n = \frac{(1.96)^2 \times 0.25 \times 0.75}{0.0004} = 1800 \text{ approximately.}$$

The next example is based on an illustration given earlier in this chapter.

### Example 13.2

It is required to estimate the average height of a large body of men. A random sample of 400 men is found to have a mean height of $69.2$ in. and a standard deviation of $2.7$ in. Find the 99 per cent confidence limits of the true average height.

The standard error of the mean height is $2.7/\sqrt{400} = 0.135$ in. From the tables, 99 per cent of the Normal Distribution lies within $2.576\sigma$ of the mean, so with 99 per cent confidence the true mean lies in the interval

$$69.2 \pm 0.135 \times 2.576 = 69.2 \pm 0.35, \text{ or } 68.85 \text{ in. to } 69.55 \text{ in.}$$

As in the previous example, the standard error is subject to sampling errors because the standard deviation could be estimated only from the sample. With small samples the uncertainty of the standard deviation means that the tables of the Normal Distribution cannot safely be used. The next two sections show how the standard deviation is estimated from a small sample and how the confidence limits have to be modified.

### Estimating the standard deviation

Let the sample consist of $n$ observations $x_1, x_2, \ldots x_n$, where $n$ is fairly small, say 25 or less. Let $\bar{x}$ be the sample mean, $\mu$ the population mean, supposed unknown, and $\sigma$ the population standard deviation, also unknown. If $\mu$ were known, $\sigma^2$ could be estimated from $1/n \, \Sigma(x - \mu)^2$. If, instead, it were estimated from $1/n \, \Sigma(x - \bar{x})^2$ the estimate would be too low, for $\Sigma(x - \bar{x})^2$ is less than $\Sigma(x - \mu)^2$. In fact, by an extension of formula (11.2),

$$\Sigma(x - \bar{x})^2 = \Sigma(x - \mu)^2 - n(\mu - \bar{x})^2$$

Now the expected value of each $(x - \mu)^2$ is $\sigma^2$, and since $\sigma_{\bar{x}}^2 = \sigma^2/n$, the expected value of $n(\mu - \bar{x})^2$ is $\sigma^2$.

$$\therefore \text{Expected value of } \Sigma(x - \bar{x})^2 = n\sigma^2 - \sigma^2$$
$$= (n - 1)\sigma^2$$

The best estimate of $\sigma^2$ is therefore $\Sigma(x - \bar{x})^2/(n - 1)$, and this statistic is generally denoted by $s^2$.

$$\therefore s = \sqrt{\frac{\Sigma(x - \bar{x})^2}{n - 1}} \tag{13.9}$$

The reader should note that some writers use $s^2$ to denote $1/n \, \Sigma(x - \bar{x})^2$, in which case the best estimate of $\sigma^2$ is $ns^2/(n - 1)$.

To recapitulate: if the 'true', or population, values of the mean and standard deviation of a distribution are unknown, they can be estimated from a sample. The sample mean ($\bar{x}$) gives an unbiased estimate of the true mean $\mu$, and the sample variance $s^2$ gives an unbiased estimate of the true variance $\sigma^2$, provided $n - 1$ is used as divisor for the latter—but not for the mean—instead of $n$.

## 'Student's $t$'

The standard error of the mean of a small sample can now be taken as $s/\sqrt{n}$, where $s$ is given by formula (13.9). Unfortunately, the factors by which this is multiplied to calculate confidence limits are also affected by the fact that $\sigma$ is not precisely known. For example, at the 95 per cent confidence level, the appropriate factor is no longer $1 \cdot 96$ but something a little greater, depending on the value of $n$. This new statistic is denoted by $t$ and is generally known as Student's $t$ after W. S. Gosset, who first described its distribution, writing under the *nom de plume* 'Student'.

The mathematical theory of the $t$-distribution is beyond the scope of this book, but a few values of $t$ are shown in Table 23 for 95 per cent and 99 per cent confidence levels. In the table, $P$ denotes the probability that a value of $t$ greater than the figure shown will occur by chance. The left-hand column represents the number of *degrees of*

**Table 23**   Selected values of *t*

| Degrees of freedom (= n − 1) | t | |
|---|---|---|
| | P = 0·05 | P = 0·01 |
| 1 | 12·71 | 63·66 |
| 2 | 4·30 | 9·92 |
| 3 | 3·18 | 5·84 |
| 5 | 2·57 | 4·03 |
| 10 | 2·23 | 3·17 |
| 20 | 2·09 | 2·85 |
| 60 | 2·00 | 2·66 |

*freedom* or independent comparisons between observations. With $n$ items there are $n - 1$ degrees of freedom, for there are $n - 1$ comparisons involving $x_1$, viz. $x_1 - x_2, x_1 - x_3, \ldots x_1 - x_n$, and any other difference between $x$'s can be expressed as the difference of two of these.

A simple example will show how $t$ can be used to calculate confidence limits.

*Example* 13.3
Six independent observers estimate the melting-point of a certain substance as 354°, 359°, 362°, 353°, 356° and 358° C respectively. Find the 99 per cent confidence limits for the true value.

The mean of the six observations is 357° C, and this is the best estimate of the true melting-point. The six deviations are (in degrees) −3, +2, +5, −4, −1 and +1.

$$\therefore s^2 = \tfrac{1}{5}\{(-3)^2 + 2^2 + 5^2 + (-4)^2 + (-1)^2 + 1^2\}$$
$$= 11\cdot2$$

$$\therefore \text{SE of mean} = \frac{s}{\sqrt{6}}$$
$$= \sqrt{\frac{11\cdot2}{6}}$$
$$= \sqrt{1\cdot8667}$$
$$= 1\cdot37$$

For five degrees of freedom and $P = 0\cdot01$, $t = 4\cdot03$, so the 99 per cent confidence limits are

$$357 \pm 1\cdot37 \times 4\cdot03 = 357 \pm 5\cdot5 \text{ approximately}$$
$$= 351\cdot5° \text{ and } 362\cdot5°$$

(Tip for examination candidates: do not waste time calculating *s* if it is not required. You still have to divide by $\sqrt{6}$. In fact, $4 \cdot 03 \times \sqrt{(11 \cdot 2/6)}$ can be worked out very quickly with a slide rule.)

## Tests of significance

Occasions sometimes arise when it is necessary to compare a statistic, such as a mean or proportion, with some specified value, or to compare two statistics with each other. For example, an experiment may be carried out to test whether a new fertilizer improves the yield of a crop, or whether a change in a chemical process increases the yield or improves the quality of the product. There will inevitably be differences in the results. The question is whether those differences are *statistically significant*, i.e. whether they indicate a real improvement in the process etc., or whether they could reasonably be attributed to chance in the form of sampling errors.

If a difference as great as, or greater than, the observed difference could only occur by chance less than once in 100 times, it is said to be significant *at the 1 per cent level*, and similarly for other *levels of significance*. A result that is significant at the 5 per cent level, i.e. that could only have occurred by chance less than 5 times (but more than once) in 100 is generally described as 'probably significant'. The 1 per cent and 5 per cent levels have no intrinsic merit, but they have proved convenient in practice and have become fixed by convention.

Putting it another way, we carry out our calculations on the assumption that the effect we are looking for does not really exist, then check to see whether the results are consistent with that assumption. Statisticians call this assumption the *Null Hypothesis*. The level of significance is the risk we are willing to take of being wrong, i.e. of rejecting the Null Hypothesis when it is true.

### Example 13.4

A plant is producing large numbers of articles of which, on average, 2 per cent are defective. In a random sample of 1000, there are 3 per cent defective. Does this indicate a significant deterioration in the process?

The expected number of defectives is 2 per cent of 1000, i.e. 20.

The standard error of the observed number is, from (13.7),

$$\sqrt{(1000 \times 0 \cdot 02 \times 0 \cdot 98)}$$
$$= \sqrt{19 \cdot 6}$$
$$= 4 \cdot 43$$

The difference between actual and expected numbers is 10, which is a little over twice the standard error. This is significant at the 5 per cent level, i.e. there is probably a real deterioration in the process.

## Single-sided and double-sided tests

In Example 13.4 the test used was *double-sided*, i.e. the probability taken as the criterion of significance was the probability of the difference from the mean of 10 or more in either direction. Sometimes, however, it is only differences in one direction that are of any interest, and in such cases a *single-sided* test is appropriate. Thus, if a new drug is being tested for a certain disease, and judged by the proportion of patients cured, the level of significance will be the probability that an increase (not merely a difference) of the magnitude found could have occurred by chance.

## Example 13.5

A process takes an average time of 35 minutes, with a standard deviation of 2·4 minutes. It is thought that a certain modification would reduce this time, and after being modified the process is repeated 12 times, giving an average time of 33·3 minutes. Is there any significant reduction in the time? (It is assumed that the variability is unchanged.)

$$\sigma = \text{SE of mean time} = \frac{2·4}{\sqrt{12}} = 0·693 \text{ min.}$$

$$x = \text{Expected mean} - \text{actual mean} = 35 - 33·3 = 1·7 \text{ min.}$$

$$\therefore \frac{x}{\sigma} = \frac{1·7}{0·693} = 2·45$$

For a single-sided test, $P < 0·01$, i.e. a difference of 2·45 or more *in the direction expected* would not occur by chance once in 100 times. It is therefore significant.

Table 24 shows the values of $x/\sigma$ (the Normal Deviate) corresponding to various values of $P$ for both single-sided and double-sided tests.

## Comparison of two estimates

The last two examples dealt with cases in which one sample statistic was compared with a known value. When two sample statistics, both

**Table 24**   Values of the Normal Deviate for single-sided and double-sided tests

| P | Single-sided | Double-sided |
|------|------|------|
| 0·1 | 1·282 | 1·645 |
| 0·05 | 1·645 | 1·960 |
| 0·02 | 2·054 | 2·326 |
| 0·01 | 2·326 | 2·576 |
| 0·001 | 3·090 | 3·291 |

subject to sampling errors, are compared with each other the problem is slightly more complicated.

We have seen that if $x$ and $y$ are any two random variables or 'statistics' with standard errors $\sigma_x$ and $\sigma_y$ respectively, the standard error $\sigma_d$ of their difference ($d = x - y$) is given by formula (13.2a),

$$\sigma_d^2 = \sigma_x^2 + \sigma_y^2$$

In the particular (and usual) case in which the expected value of $d$ is zero and we are testing to see whether the actual difference between two sample means or proportions could have occurred by chance, we calculate $\sigma_d$, divide it into $d$, and compare the results with the appropriate table.

### (a) Comparison of two sample proportions

Suppose two samples of $n_1$ and $n_2$ items are being compared in respect of a certain attribute, and the proportions of successes in those samples are $p_1$ and $p_2$ respectively. Then the total number of successes in the combined sample is $n_1 p_1 + n_2 p_2$, and the overall proportion of successes is given by

$$p = \frac{n_1 p_1 + n_2 p_2}{n_1 + n_2}$$

On the assumption that the two samples were taken at random from the same population (i.e. on the Null Hypothesis), this value $p$ is the best estimate of the true proportion in the population. The standard error of the observed proportion in a sample of $n$ items is, therefore, from (13.8),

$$\sqrt{\frac{pq}{n}}, \text{ where } q = 1 - p$$

So, from (13.2a) and (13.7), the standard error $\sigma_d$ of $p_1 - p_2$ is given by

$$\sigma_d{}^2 = \sigma_{p_1}{}^2 + \sigma_{p_2}{}^2 = \left(\frac{1}{n_1} + \frac{1}{n_2}\right) pq \qquad (13.10)$$

Then $p_1 - p_2$ can be compared with $\sigma_d$ by a single-sided or double-sided test, whichever is appropriate.

*Example* 13.6

Out of 200 articles made on one machine, 22 per cent are faulty; out of 120 made on another machine, 30 per cent are faulty. Is there a significant difference between the two machines? (Note the assumption that if the difference is significant, it is due to the machines. Such assumptions are often implied in questions without being stated.)

The proportion of faulty articles in the combined batch of 320 is

$$\frac{44 + 36}{320} = \frac{1}{4}$$

So, from (13.10),

$$\sigma_d{}^2 = \left(\frac{1}{200} + \frac{1}{120}\right) \times \frac{1}{4} \times \frac{3}{4}$$

$$= \frac{1}{400}$$

$$\therefore \sigma_d = \frac{1}{20}, \qquad \text{or 5 per cent.}$$

The actual difference of 8 per cent between the two samples is only $1 \cdot 6$ times its standard error, and a difference as great as this could occur by chance at least 1 in 10 times. The verdict is therefore 'not proven'. Note that we have not proved either that there is a real difference or that there is not. Probably, further experiment is indicated.

### (b) Comparison of two sample means

Suppose we wish to compare the means $\bar{x}_1$ and $\bar{x}_2$ of two samples, of size $n_1$ and $n_2$ items respectively, to check whether they could reasonably have come from the same distribution. Let us suppose first that the standard deviation $\sigma$ of that distribution is known.

From formula (13.6), the standard errors of $\bar{x}_1$ and $\bar{x}_2$ are respectively $\sigma/\sqrt{n_1}$ and $\sigma/\sqrt{n_2}$, and from formula (11.2a), the standard error of $\bar{x}_1 - \bar{x}_2$ is given by

$$\sigma_d{}^2 = \frac{\sigma^2}{n_1} + \frac{\sigma^2}{n_2} = \left(\frac{1}{n_1} + \frac{1}{n_2}\right) \sigma^2 \qquad (13.11)$$

*Example* 13.7

Two machines both produce articles varying in weight with a standard deviation of $1.5$ g; one turns out 50 with a mean weight of $172.1$ g and the other turns out 30 with a mean weight of $172.9$ g. Is one machine biased with respect to the other?

The standard error of the difference is, from (13.11),

$$\sigma_d = \sqrt{\left(\frac{1}{50} + \frac{1}{30}\right)} \times 1.5$$
$$= 0.346$$

The actual difference of $0.8$ g is $2.31$ times its standard error, and the table of the Normal distribution shows that a difference as great as this could only occur by chance about once in fifty times. We conclude that one of the machines is probably biased.

If $\sigma$ is not known, it must be estimated by pooling the sums of squares from the two samples. Strictly speaking, the two estimates of $\sigma^2$ should be compared for consistency, but that takes us beyond the scope of this chapter.

From first sample, expected value of

$$\Sigma(x - \bar{x}_1)^2 = (n_1 - 1)\sigma^2$$

and from second sample, expected value of

$$\Sigma(x - \bar{x}_2)^2 = (n_2 - 1)\sigma^2$$

The best estimate of $\sigma^2$ is therefore $s^2$, where

$$(n_1 + n_2 - 2)s^2 = \Sigma(x - \bar{x}_1)^2 + \Sigma(x - \bar{x}_2)^2$$

The standard error of $d(=\bar{x}_1 - \bar{x}_2)$ is then given by

$$s_d^2 = \left(\frac{1}{n_1} + \frac{1}{n_2}\right) s^2 \tag{13.12}$$

If $n_1$ and $n_2$ are small, $s^2$ is uncertain and we must again use $t$ instead of the Normal deviate. Since $s^2$ has been obtained with $n_1 + n_2 - 2$ degrees of freedom, this is the appropriate number to use with $t$.

*Example* 13.8

Seven eight-week-old chickens reared on a special diet weigh 13, 15, 13, 18, 15, 15 and 16 oz, and five chickens of similar age fed on ordinary diet weigh 9, 11, 15, 11 and 14 oz. Is there sufficient evidence that the special diet has caused an increase in weight?

Denoting the two groups by suffixes 1 and 2 respectively, it is easily seen that

$$\bar{x}_1 = 15, \qquad \bar{x}_2 = 12, \qquad d = 3$$
$$\Sigma(x - \bar{x}_1)^2 = 18, \qquad \Sigma(x - \bar{x}_2)^2 = 24$$

$$\therefore 10s^2 = 18 + 24 = 42$$
$$s^2 = 4.2$$

$$\therefore \quad s_d^2 = (\tfrac{1}{7} + \tfrac{1}{5})4.2 = 1.44$$
$$s_d = 1.2$$

$$\therefore \quad t = \frac{3}{1.2} = 2.5$$

Since $s^2$ has been calculated with 10 degrees of freedom, we look up the corresponding values of $t$ in Table 23 (page 151) and find that a value as great as 2.23 would be significant at the 5 per cent level. But be careful: that is for a double-sided test. We are only interested in differences in one direction. An *increase* in average weight of the amount found is significant at the 2.5 per cent level, i.e. it would occur by chance less than once in forty times. We conclude that the effect is probably significant. As it happens, we should have done so if the test had been double-sided, but with rather less confidence.

## Paired comparisons

It is often possible to increase the sensitivity of a test by removing most of the variability from our data. In Example 13.8, for instance, we might have reduced our sum of squares and consequently our value of $s^2$ by pairing off chickens from the same brood and allocating one to the special diet and the other to the ordinary diet. The allocation must, of course, be done by some random process.

## Example 13.9

In order to test a possible treatment for increasing the abrasion resistance of rubber, eight test-pieces are taken from different sheets and each test-piece is divided into two parts, one of which receives the treatment. All sixteen pieces are tested, with the following results:

| Test-piece | Treated | Untreated |
|:---:|:---:|:---:|
| 1 | 15 | 12 |
| 2 | 14 | 11 |
| 3 | 13 | 13 |
| 4 | 16 | 14 |
| 5 | 12 | 11 |
| 6 | 12 | 10 |
| 7 | 15 | 14 |
| 8 | 11 | 11 |

If these data were analysed as in Example 13.8, we should obtain $t = 1 \cdot 82$ for 14 degrees of freedom, which is barely significant at the 5 per cent level. However, there is considerable variation between test-pieces, which can be removed by taking the differences between corresponding halves and analysing the eight figures so obtained.

| Test-piece | Difference $(x)$ |
|:---:|:---:|
| 1 | 3 |
| 2 | 3 |
| 3 | 0 |
| 4 | 2 |
| 5 | 1 |
| 6 | 2 |
| 7 | 1 |
| 8 | 0 |

If $x$ denotes the result of subtracting the 'untreated' from the corresponding 'treated' observation, we find that

$$\bar{x} = 1 \cdot 5, \qquad \Sigma(x - \bar{x})^2 = 10$$

$$\therefore s^2 = \frac{10}{7}$$

and SE of $\bar{x}$

$$= \frac{s}{\sqrt{8}} = \sqrt{\frac{10}{56}}$$

Since we are testing whether $\bar{x}$ is significantly greater than zero,

$$t = 1 \cdot 5 \div \sqrt{\frac{10}{56}} = 3 \cdot 55.$$

For 7 degrees of freedom, this is significant at the 1 per cent level—in fact it would be, even for a double-sided test. Thus, what we have lost in degrees of freedom we have more than gained in sensi-tivity, by reducing the standard error.

## Distribution-free tests

No modern book on statistics is complete without at least a brief reference to *distribution-free* or *non-parametric tests*. The two terms are often treated as synonymous, but this is not strictly correct. 'Distribution-free' means that the test does not depend on the form of distribution; a non-parametric test may (but seldom does) assume a particular form of distribution but makes no assumption about the SD or other parameters.

Most of the standard tests postulate a Normal parent distribution. With some tests, notably the $t$-test, a moderate departure from Nor-mality makes little difference to their validity; such tests are said to be *robust*. Even with Normal data, however, it may be much quicker, and adequate for the purpose in hand, to use a distribution-free test. In Example 13.8, instead of $t$ we could use *Lord's Test* for the difference between two sample means, using the formula

$$t' = \frac{|x_1 - x_2|}{w_1 + w_2} \tag{13.13}$$

where $w_1$ and $w_2$ are the ranges of the two samples. In this case,

$$t' = \frac{15 - 12}{5 + 6} = 0 \cdot 273.$$

From the tables (if you can find them), this is significant at the 5 per cent level.

In many distribution-free tests, measurements are replaced by ranks $(1, 2, \ldots n)$ or simply by division into two categories as in the *Sign Test*, which is a straightforward comparison of the numbers of positive and negative observations, using the binomial distribution with $p = q = \frac{1}{2}$. In Example 13.9, for instance, we have six positive and no negative differences, ignoring the zeros. The probability of all six being positive by chance is $1/2^6$, or $1/64$ (remember, this was a single-sided test). The results of the experiment are therefore signifi-cant at the 2 per cent level.

Distribution-free tests are useful with mixed or erratic distributions, and with data that can be put in rank order but not measured, but they have the following drawbacks:

(i) They can seldom be generalized to provide confidence limits.

(ii) Some of the formulae are extremely difficult to prove or to remember. One well-known book takes four pages to prove the formula for the variance in the Wald-Wolfowitz Runs Test.

(iii) They usually require special tables that are not readily available.

### Concluding remarks

Only the fringe of this subject can be touched in an elementary text-book, and it has only been possible to deal cursorily with the treatment of small samples. The reader who is sufficiently interested and equipped to read further should refer to more advanced books. Certain aspects of sampling methods will be discussed further in Chapter 16.

It only remains to sound a note of caution concerning tests of significance. The conventional 1 per cent and 5 per cent are useful so long as they are treated with discretion. An experimental result significant even at the 10 per cent level may justify changing a process if it supports an opinion already held for good reasons. On the other hand, the management will need something more convincing than a result that is 'probably significant' before embarking on a large capital construction programme. Indeed, they will need to be satisfied that the improvement is great enough to justify the change.

This raises a point of great importance. A test of significance often tells nothing about the size of the effect being tested. A difference may be statistically significant but in practice insignificant. Confidence intervals are often more informative than tests of significance, because they show the probable size of the difference being investigated and at the same time indicate the degree of uncertainty attached to it. Indeed, we may know at the outset that there must be a difference. The only question is how big it is.

Finally, do remember that the Null Hypothesis can never be proved. There may be a difference so small that one would need a very large sample to detect it. If a test is not significant, the Null Hypothesis may be accepted, but that does not mean it has been proved.

## Exercises

13.1   Give formulae for the following, explaining your notation and stating what assumptions have to be made in calculating and in interpreting them:

(a) Standard error of a mean.
(b) Standard error of a proportion.
(c) Standard error of a number of accidents.
(d) Standard error of a sum of two readings from the same population.
(e) Standard error of the difference between two readings from the same population.                                                        (*IoS*)

13.2   In order to find whether intensive national advertising of milk affected sales in a particular area, two systematic samples of families were taken from the milkroundsmen's books, one before and one after the campaign. With the help of means and standard deviations, determine whether the campaign was successful, ignoring all other factors influencing the pattern of milk sales.

| Number of pints bought per week | Number of families in | |
|---|---|---|
| | Sample 1 Before the campaign | Sample 2 After the campaign |
| Up to 5 | 77 | 81 |
| $5\frac{1}{2}$–$10\frac{1}{2}$ | 151 | 128 |
| 11 –16 | 213 | 194 |
| $16\frac{1}{2}$–$21\frac{1}{2}$ | 254 | 295 |
| 22 –27 | 428 | 372 |
| $27\frac{1}{2}$–$32\frac{1}{2}$ | 230 | 323 |
| 33 –38 | 92 | 146 |
| $38\frac{1}{2}$ and over | 45 | 79 |

(*ICMA*)

13.3   After corrosion tests, 42 of 536 metal components treated with Primer A and 91 of 759 components treated with Primer B showed signs of rusting. Test the hypothesis that Primer A is superior to Primer B as a rust inhibitor.                                    (*ICMA*)

13.4   What is meant by the term 'confidence limit'?
In a random sample of Manchester residents 410 people, out of 1000 interviewed, looked at a certain television programme. Find the 95 per

cent confidence limits for the percentage of all the residents who looked at the programme. If a sample of 500 Londoners showed that 52 per cent looked at the same programme, would you consider this evidence of a difference in viewing habits between the two cities?

(*IoS*)

13.5 In a sample of 500 garages it was found that 170 sold tyres at prices below those recommended by the manufacturer.

(*a*) Estimate the percentage of all garages selling below list price.

(*b*) Calculate the 95 per cent confidence limits for this estimate and explain briefly what these mean.

(*c*) What size sample would have to be taken in order to estimate the percentage to within 2 per cent? (*ICMA*)

13.6 A certain large country has two political parties, 'Blues' and 'Greens'. Approximately 55 per cent of the electorate vote Blue and 45 per cent vote Green.

(*a*) It is required to estimate, with 95 per cent confidence, percentages within $\pm 1$ unit. Find the sample size needed.

(*b*) Because of known differences in their traditional voting habits, you are now told that 75 per cent of the electorate is wage-earning and 25 per cent salaried or self-employed. In a sample of the size calculated in (*a*) above, only 72 per cent of the persons interviewed were found to be wage-earning. To what extent would this affect any conclusions drawn from the sample? What remedy would you propose for such a situation? (*IoS*)

13.7 Seeds of a certain vegetable are required by law to have a minimum germination rate of 80 per cent when they are sold. A seedsman performs a trial of 50 seeds and observes that 45 germinate. Is he justified in believing that his seeds are up to standard?

An inspector tests a further random sample of 50 seeds and finds that only 32 germinate. Is this sufficient evidence that the stock is substandard? Are the inspector's results consistent with those of the seedsman? (*IoS*)

13.8 Explain what is meant by the term 'confidence interval'.

A sample of 12 observations from a certain population gave the following total and sum of squares:

$$\Sigma x = 24, \qquad \Sigma x^2 = 92.$$

Derive a 95 per cent confidence interval for the population mean.

At a later date a further sample of 15 observations gave the total and sum of squares:

$$\Sigma y = 45, \qquad \Sigma y^2 = 241.$$

Stating any assumptions you make, test at the 5 per cent level of significance whether the population mean has changed.          (*IoS*)

13.9   The time taken for a maintenance job on part of a textile process has been established to have a mean of 3 hrs 24 mins and a standard deviation of 12 minutes. How many tests would be required by a new method of maintenance to detect a difference of 9 minutes in average time for the job with 95 per cent confidence?

The following eight times were recorded by a new method:

|  |  |
|---|---|
| 3 hrs   6 mins | 3 hrs 12 mins |
| 2 hrs 54 mins | 3 hrs   0 mins |
| 3 hrs 18 mins | 3 hrs 24 mins |
| 3 hrs 30 mins | 3 hrs 12 mins |

Obtain an estimate of the mean of the new method and calculate its 95 per cent confidence limits.

Is there any indication that the average time for the new method differs from the average time for the established method?          (*IoS*)

13.10   A large number of employees are engaged in an office on similar routine tasks. It has been suggested that a change in the equipment used will improve productivity. The efficiency of ten employees selected at random is measured with the old and the new equipment. The following data are then available:

| Employee | Efficiency rating | |
|---|---|---|
|  | Old equipment | New equipment |
| A | 60 | 60 |
| B | 64 | 67 |
| C | 56 | 60 |
| D | 45 | 48 |
| E | 96 | 94 |
| F | 64 | 64 |
| G | 54 | 59 |
| H | 40 | 50 |
| I | 78 | 78 |
| J | 75 | 77 |

Do the data provide evidence of the superiority of the new equipment? (*IoS*)

13.11 In a clinical comparison of the effectiveness of two aerosols, *A* and *B*, in the relief of bronchial asthma, ten patients were selected. Each patient was given *A* on one occasion and *B* on a different occasion, the order of administration being decided by a random procedure. On each occasion the patient's forced expiratory volume (FEV) was estimated before treatment and at intervals up to one hour after treatment. The criterion of effectiveness was taken to be the maximum percentage increase in FEV compared with the pretreatment readings:

| Patient | Maximum percentage increase in FEV after | |
|---|---|---|
| | Aerosol A | Aerosol B |
| 1 | 28 | 24 |
| 2 | 22 | 16 |
| 3 | 10 | 5 |
| 4 | 40 | 17 |
| 5 | 18 | 23 |
| 6 | 57 | 52 |
| 7 | 49 | 30 |
| 8 | 40 | 16 |
| 9 | 34 | 14 |
| 10 | 63 | 31 |

Is there any convincing evidence of a difference between the two aerosols? (*IoS*)

# 14 Regression and correlation

The previous few chapters have been concerned with the distribution of a single variable; the present chapter deals with the interrelations between two variables.

## The meaning of regression

In the exact sciences like astronomy, chemistry and physics, it is possible to formulate laws connecting several quantities, e.g. the density, temperature and pressure of a gas, so that any one of these quantities can be determined (subject only to small experimental errors) from the others. In meteorology, prediction is much less certain, but it is still possible to forecast the weather with considerable success. In the living sciences like biology and agriculture, there are so many unknown factors at work that very different results may be obtained from what appear to be identical causes.

Even here, however, the law of averages holds good, e.g. big men and women tend to have big children, although the association is only partial. The adult sons of men below average height will also tend to be short, but on the whole they will be nearer the average height than their fathers, e.g. the sons of men 5 ft tall might, on the average, reach about 5 ft 4 in. It was this 'regression' to the normal, noticed by Sir Francis Galton in his research on heredity, that gave the name 'regression analysis' to this branch of statistics.

While, therefore, the son's height cannot be deduced from the father's, it is possible, knowing the father's height, to make a better forecast of the son's height than by simply taking the average of the population. In Chapter 13 it was shown how to estimate a parameter and find the precision of that estimate. In this chapter it will be shown how to find the best estimate of one variable partly dependent on another, knowing the value of the other; how to find the precision of the estimate is beyond the scope of this book.

## Correlation

Two or more variables that tend to move in sympathy are said to be *correlated*. In an elementary textbook it is possible to consider only two variables. When high values of one are associated with high values of the other, as in the case of fathers' and sons' heights, they are said to be *directly* or *positively correlated*. When high values of one tend to accompany low values of the other, e.g. unemployment and labour turnover, they are *inversely* or *negatively correlated*.

There are, of course, degrees of correlation, i.e. two variables may be highly correlated or only slightly correlated. Thus, candidates' examination marks in two mathematics papers are likely to be highly correlated, but their marks in mathematics and art or woodwork will show little if any correlation. It will be shown later how the degree of correlation between two variables can be measured.

## Scatter diagrams

The existence of correlation can be shown graphically by means of a *scatter diagram*, in which each point corresponds to a pair of observations, one variable being plotted horizontally and the other vertical-

**Table 25** Average duration of absence due to injury in various factories in 1975 and 1976

| Factory | 1975 | 1976 |
|---------|------|------|
|         | Hours | Hours |
| A | 232 | 205 |
| B | 206 | 175 |
| C | 252 | 273 |
| D | 101 | 122 |
| E | 154 | 176 |
| F | 224 | 236 |
| G | 158 | 217 |
| H | 321 | 324 |
| I | 153 | 151 |
| J | 74 | 46 |
| K | 154 | 109 |
| L | 170 | 183 |
| M | 303 | 226 |

ly. Table 25 shows hypothetical data collected for different factories on the average duration of absence (in working hours) due to industrial injuries in successive years. Clearly there is wide variation between factories, but the scatter diagram shown in Fig. 40 suggests that these averages are fairly consistent from year to year. From this one might infer that the severity of the injury depends largely on the type of work done in each factory.

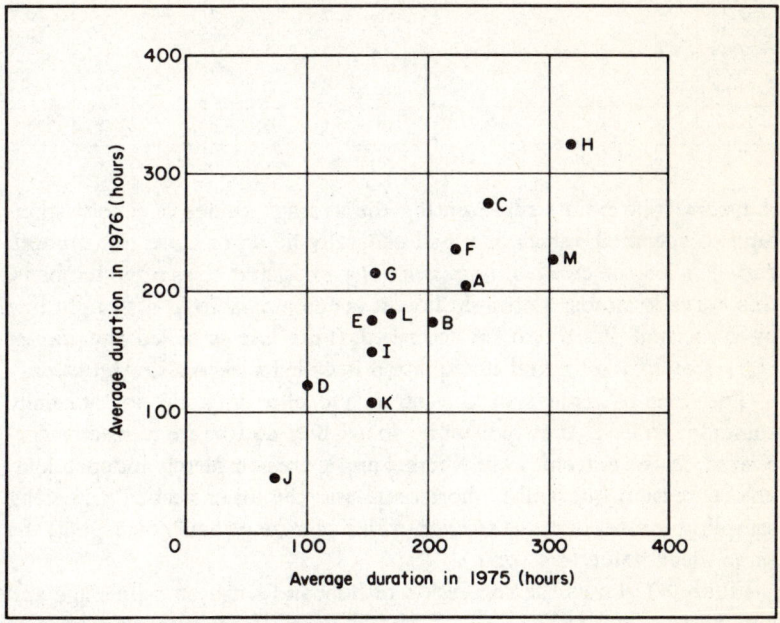

**Fig. 40** Scatter diagram (from Table 25)

## Curves and lines of regression

When there are many observations, the labour of plotting a scatter diagram becomes prohibitive, and it is better to compile a *correlation table*, which may be described as a two-way frequency distribution. An example of this is shown in Table 26, which has been adapted from a table in the Registrar-General's *Statistical Review for 1951*, the frequency of each 'cell' being divided by 1000 and rounded off to the nearest integer.

For each value or class-interval of one variable, the average value of the other variable may be calculated and plotted on a *regression*

**Table 26** Correlation table showing combined ages at marriage of 353 married couples

Ages of husbands

| Ages of wives | 16–19 | 20–24 | 25–29 | 30–34 | 35–39 | 40–44 | 45–49 | 50–54 | 55–59 | 60–64 | 65–69 | Total |
|---|---|---|---|---|---|---|---|---|---|---|---|---|
| 16–19 | 6 | 37 | 9 | 1 | — | — | — | — | — | — | — | 53 |
| 20–24 | 3 | 95 | 57 | 13 | 3 | 1 | — | — | — | — | — | 172 |
| 25–29 | — | 11 | 28 | 14 | 6 | 2 | 1 | — | — | — | — | 62 |
| 30–34 | — | 1 | 6 | 8 | 5 | 3 | 1 | 1 | — | — | — | 25 |
| 35–39 | — | — | 1 | 3 | 4 | 3 | 2 | 1 | — | — | — | 14 |
| 40–44 | — | — | — | 1 | 2 | 3 | 2 | 1 | 1 | — | — | 10 |
| 45–49 | — | — | — | — | 1 | 1 | 2 | 2 | 1 | 1 | — | 8 |
| 50–54 | — | — | — | — | — | — | 1 | 1 | 1 | 1 | — | 4 |
| 55–59 | — | — | — | — | — | — | — | — | 1 | 1 | 1 | 3 |
| 60–64 | — | — | — | — | — | — | — | — | — | 1 | 1 | 2 |
| Total | 9 | 144 | 101 | 40 | 21 | 13 | 9 | 6 | 4 | 4 | 2 | 353 |

*diagram*. The points representing the average values of $y$ corresponding to specified values of $x$ will generally lie on or close to a smooth curve, called the *curve of regression of $y$ on $x$*, and if, as often happens, this curve is almost a straight line, it is customary to fit a straight line by a method shortly to be described. This line is called the *line of regression of $y$ on $x$* and its equation is called a *regression equation*.

The lines of regression of $y$ on $x$, and of $x$ on $y$, do not normally coincide; indeed, they will only do so if $x$ and $y$ are perfectly correlated. In the extreme case where $x$ and $y$ are completely independent, one regression line will be horizontal and the other vertical (ignoring sampling errors) because to every value of $x$ (or $y$) will correspond the same mean value of $y$ (or $x$).

Figure 41 shows the regression of husband's age on wife's age and vice versa. The broken lines $A$ and $B$ would form smooth curves of regression if based on very large numbers of observations. Obviously there is a high correlation between the ages of husband and wife, as would be expected.

As a rule only one regression line or curve is required, because one variable $y$ is generally regarded as dependent on the other variable $x$.

## Regression equations: the method of least squares

It will be assumed in this section that the regression of $y$ on $x$ is linear, i.e. that it is sufficient to fit a straight line to the data. The same principles, however, can be extended to parabolas and other curves. An exponential curve can be fitted by finding the line of regression of log $y$

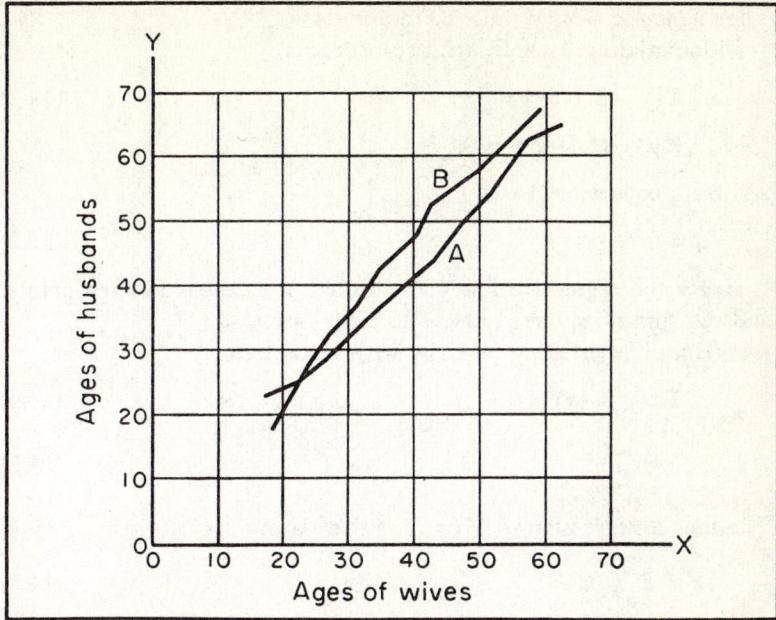

**Fig. 41** Regression diagram illustrating regression of husband's age on wife's age $(A)$ and vice versa $(B)$ (from Table 26)

on $x$ and then taking anti-logs. The non-mathematically-minded student may omit the part between Equations (14.1) and (14.5).

Suppose there are $n$ pairs of items

$$(x_1, y_1), (x_2, y_2) \ldots (x_n, y_n),$$

and that it is required to find the regression line of $y$ on $x$.

It will save confusion if $X$ and $Y$ represent the variables and $x$ and $y$ denote observed values of $X$ and $Y$. Let the required line be $Y = a + bX$, where $a$ and $b$ are constants to be determined. Then any observed value of $Y$, say $y_r$, may be regarded as the sum of (i) the estimate from the regression equation, viz. $a + bx_r$, and (ii) a residual $y_r - a - bx_r$. The Method of Least Squares consists in finding the values of the *regression coefficients* $a$ and $b$ that minimize the sum of squares of the residuals. This is equivalent to minimizing the standard error of an estimated value of $y$.

To simplify the mathematics, suppose both variables are measured from their means, so that

$$\Sigma x = \Sigma y = 0 \tag{14.1}$$

Then $\Sigma(y - a - bx)^2$ must be minimum.

Differentiating partially with respect to $a$,

$$\Sigma(y - a - bx) = 0 \qquad (14.2)$$

$$\therefore \Sigma y - na - b\Sigma x = 0,$$

and from equations (14.1),

$$a = 0. \qquad (14.3)$$

Hence the regression line goes through the centroid of the points, and the sum of squares reduces to $\Sigma(y - bx)^2$.

Again, differentiating partially with respect to $b$,

$$\Sigma x(y - bx) = 0 \qquad (14.4)$$

$$\therefore b = \frac{\Sigma xy}{\Sigma x^2} \qquad (14.5)$$

The line of regression of $Y$ on $X$ is therefore:

$$Y = \frac{\Sigma xy}{\Sigma x^2} X, \qquad (14.6)$$

or, if $X$ and $Y$ are not measured from the means,

$$Y - \bar{y} = \frac{\Sigma(x - \bar{x})(y - \bar{y})}{\Sigma(x - \bar{x})^2}(X - \bar{x}) \qquad (14.7)$$

Similarly, the line of regression of $X$ on $Y$ is

$$X - \bar{x} = \frac{\Sigma(x - \bar{x})(y - \bar{y})}{\Sigma(y - \bar{y})^2}(Y - \bar{y}) \qquad (14.8)$$

If $\bar{x}$ and $\bar{y}$ are not convenient quantities, a short-cut method may be used for calculating the numerator of the regression coefficients similar to that found in Chapter 11 for $\Sigma(x - \bar{x})^2$.

$$\begin{aligned}
\Sigma(x - \bar{x})(y - \bar{y}) &= \Sigma xy - \bar{x}\Sigma y - \bar{y}\Sigma x + n\bar{x}\bar{y} \\
&= \Sigma xy - \bar{x}(n\bar{y}) - \bar{y}(n\bar{x}) + n\bar{x}\bar{y} \\
&= \Sigma xy - n\bar{x}\bar{y} \\
&= \Sigma xy - \frac{1}{n}(\Sigma x)(\Sigma y) \qquad (14.9)
\end{aligned}$$

*Example* 14.1

Find the line of regression of $Y$ (cinema admissions per head) on $X$ (television licences per 100 population) from the following data for ten towns, covering one year.

| Town | TV licences per 100 of population | Cinema admissions per head of population |
|------|-----------------------------------|------------------------------------------|
| A | 12 | 12·8 |
| B | 39 | 10·9 |
| C | 16 | 13·7 |
| D | 30 | 12·4 |
| E | 43 | 8·8 |
| F | 17 | 13·3 |
| G | 47 | 10·1 |
| H | 15 | 11·6 |
| I | 34 | 10·6 |
| J | 22 | 10·4 |

Take trial means of 27 for $X$ and 11 for $Y$ and to simplify the arithmetic let

$$x = X - 27, \qquad y = 10(Y - 11)$$

The calculation is as follows:

| Town | $x$ | $y$ | $x^2$ | $y^2$ | $xy$ | $x + y$ | $(x + y)^2$ |
|------|-----|-----|-------|-------|------|---------|-------------|
| A | −15 | +18 | 225 | 324 | −270 | + 3 | 9 |
| B | +12 | − 1 | 144 | 1 | − 12 | +11 | 121 |
| C | −11 | +27 | 121 | 729 | −297 | +16 | 256 |
| D | + 3 | +14 | 9 | 196 | + 42 | +17 | 289 |
| E | +16 | −22 | 256 | 484 | −352 | − 6 | 36 |
| F | −10 | +23 | 100 | 529 | −230 | +13 | 169 |
| G | +20 | − 9 | 400 | 81 | −180 | +11 | 121 |
| H | −12 | + 6 | 144 | 36 | − 72 | − 6 | 36 |
| I | + 7 | − 4 | 49 | 16 | − 28 | + 3 | 9 |
| J | − 5 | − 6 | 25 | 36 | + 30 | −11 | 121 |
| Total | + 5 | +46 | 1473 | 2432 | −1369 | +51 | 1167 |

The last two columns are inserted as a useful check, since

$$\Sigma x + \Sigma y = \Sigma(x + y)$$

and

$$\Sigma x^2 + \Sigma y^2 + 2\Sigma xy = \Sigma(x + y)^2$$

Substituting,

$$5 + 46 = 51$$

and

$$1473 + 2432 + 2(-1369) = 1167$$

which confirms our arithmetic.

Then

$$\bar{x} = +0 \cdot 5, \bar{y} = +4 \cdot 6$$

$$\Sigma(x - \bar{x})^2 = 1473 - \frac{(+5)^2}{10} = 1470 \cdot 5$$

$$\Sigma(y - \bar{y})^2 = 2432 - \frac{(+46)^2}{10} = 2220 \cdot 4$$

$$\Sigma(x - \bar{x})(y - \bar{y}) = -1369 - \frac{(+5)(+46)}{10} = -1392$$

So the line of regression of $y$ on $x$ is

$$y - 4 \cdot 6 = -\frac{1392}{1470 \cdot 5}(x - 0 \cdot 5)$$

or, transforming to the original variables

$$Y - 11 \cdot 46 = -\frac{1392}{14,705}(X - 27 \cdot 5)$$

which reduces to

$$Y = -0 \cdot 0947X + 14 \cdot 06$$

Figure 42 shows the ten points and the regression line.

**Fig. 42**   Regression diagram illustrating Example 14.1

## Coefficient of correlation

One object in fitting a regression line was to enable improved estimates to be given of the values of $y$ corresponding to specified values of $x$, i.e. to reduce the standard error of such estimates. A comparison of $\Sigma y^2$ with $\Sigma(y - bx)^2$, where $x$ and $y$ are measured from their means and $b$ is given by Equation (14.5), will show how far this object has been achieved.

$$
\begin{aligned}
\Sigma(y - bx)^2 &= \Sigma y^2 - 2b\,\Sigma xy + b^2\,\Sigma x^2 \\
&= \Sigma y^2 - 2\left(\frac{\Sigma xy}{\Sigma x^2}\right)(\Sigma xy) + \left(\frac{\Sigma xy}{\Sigma x^2}\right)^2 (\Sigma x^2) \\
&= \Sigma y^2 - \frac{(\Sigma xy)^2}{\Sigma x^2} \\
&= (1 - r^2)\,\Sigma y^2, \tag{14.10}
\end{aligned}
$$

where

$$
r = \frac{\Sigma xy}{\sqrt{[(\Sigma x^2)(\Sigma y^2)]}} \tag{14.11}
$$

This quantity $r$ is called the *coefficient of correlation* between $x$ and $y$. It is a suitable measure of correlation for the following reasons:

1. If there is no real correlation, the average value of $y$ (ignoring sampling errors) is zero for every value of $x$, and $\Sigma xy = 0$, consequently $r = 0$.

2. If there is perfect linear correlation, every residual is zero, and from Equation (14.10), $r = \pm 1$. It is easily seen that $r = +1$ when correlation is direct and $-1$ when it is inverse.

3. Since $\Sigma(y - bx)^2$ cannot be negative, $r$ cannot be numerically greater than 1.

Reverting to the original co-ordinates, Formula (14.11) becomes:

$$
r = \frac{\Sigma(x - \bar{x})(y - \bar{y})}{\sqrt{[\{\Sigma(x - \bar{x})^2\}\{\Sigma(y - \bar{y})^2\}]}} \tag{14.11a}
$$

*Example* 14.2

Find the coefficient of correlation between cinema admissions and TV licences in the data of Example 14.1.

The requisite sums of squares and products have already been calculated—in fact $\Sigma(y - \bar{y})^2$ was not needed for the regression line, but was included for convenience. From (14.11a),

$$
r = \frac{-1392}{\sqrt{(1470{\cdot}5 \times 2220{\cdot}4)}} = -0{\cdot}77
$$

In practice, we should hesitate to compute either the correlation coefficient or a regression line with only ten points. There does, however, seem to be a fairly strong negative correlation, and the regression line might help us to predict the cinema attendance of any town if we knew the number of TV licences and the size of population, provided we keep within the range of our observations. The regression line might not hold, for example, for values of $X$ greater than 50.

The regression coefficients can be expressed simply by means of $r$. Putting

$$\Sigma(x - \bar{x})^2 = n\sigma_x^2 \text{ and } \Sigma(y - \bar{y})^2 = n\sigma_y^2,$$

$$r = \frac{\Sigma(x - \bar{x})(y - \bar{y})}{n\sigma_x\sigma_y} \tag{14.12}$$

and

$$b = \frac{\Sigma(x - \bar{x})(y - \bar{y})}{n\sigma_x^2} = r\frac{\sigma_y}{\sigma_x} \tag{14.13}$$

Hence the line of regression of $y$ on $x$ is:

$$Y - \bar{y} = r\frac{\sigma_y}{\sigma_x}(X - \bar{x}) \tag{14.14}$$

Similarly, the line of regression of $x$ on $y$ is

$$X - \bar{x} = r\frac{\sigma_x}{\sigma_y}(Y - \bar{y}) \tag{14.15}$$

### Significance of the correlation coefficient

Like other parameters, the correlation coefficient is subject to sampling errors, and it is customary to denote by $\rho$ the true value, generally unknown, of which $r$ is an estimate. The distribution of $r$ is extremely complicated and an adequate treatment of it cannot be attempted here. Even when $\rho = 0$, i.e. when there is no real correlation, the distribution of $r$ is by no means simple, but for large values of $n$ the standard error of $r$ may be taken as $1/\sqrt{n}$. Thus, if $n = 100$, an observed value of $0 \cdot 2$ is probably significant and one of $0 \cdot 3$ almost certainly significant.

Great care must be observed in judging whether or how far the correlation is significant in the everyday sense of the word. In the first place, it only measures linear correlation and is not suitable for data which, when plotted, form a parabola or similar curve. In the second place, correlation does not necessarily mean cause and effect. It is quite easy to find 'nonsense correlations' between things that have no connection whatever, except that they are both correlated with time or some other variable. Thus a rising divorce rate might be correlated with a falling mortality rate, but that would not mean that one caused the other. Again, had we taken total cinema admissions and total number of TV licences in Example 14.1 instead of rates per head of population, we should have found them positively, not negatively, correlated because both would have been correlated with size of population.

Correlation coefficients, like averages, are of little value with very mixed data. It is nearly always worth while to plot the data to make sure they are homogeneous. Figure 43 shows the kind of thing that can happen.

**Fig. 43** Scatter diagram of data unsuitable for correlation coefficient

## Rank correlation

It sometimes happens that the characteristics whose possible association is being investigated cannot be measured. Thus, a schoolmaster may want to know how far a boy's ability in games determines his popularity, but it would be extremely difficult to give either of these qualities a numerical value. However, it is fairly easy to place the boys in order for each of these qualities and to measure the agreement between the two rankings by means of a *rank correlation coefficient*. There are two in common use, one due to Kendall, the other to Spearman. The latter is described here.

The principle is the same as before, but the calculation is simplified because both $x$ and $y$ take the integral values 1, 2, ... $n$.

$$\bar{x} = \bar{y} = \frac{n + 1}{2}$$

$$\Sigma(x - \bar{x})^2 = \Sigma_1^n x^2 - n\bar{x}^2$$

$$= \frac{n(n + 1)(2n + 1)}{6} - n\left(\frac{n + 1}{2}\right)^2$$

$$= \frac{n(n^2 - 1)}{12}$$

and similarly,

$$\Sigma(y - \bar{y})^2 = \frac{n(n^2 - 1)}{12}$$

$$\therefore r = \frac{12\Sigma(x - \bar{x})(y - \bar{y})}{n(n^2 - 1)}$$

This formula can be further simplified, since

$$2(x - \bar{x})(y - \bar{y}) = (x - \bar{x})^2 + (y - \bar{y})^2 - \{(x - \bar{x}) - (y - \bar{y})\}^2$$

$$= (x - \bar{x})^2 + (y - \bar{y})^2 - (x - y)^2$$

Putting    $d = x - y$,

$$2\Sigma(x - \bar{x})(y - \bar{y}) = \Sigma(x - \bar{x})^2 + \Sigma(y - \bar{y})^2 - \Sigma d^2$$

$$= \frac{n(n^2 - 1)}{6} - \Sigma d^2$$

$$\therefore r = \frac{n(n^2 - 1) - 6\Sigma d^2}{n(n^2 - 1)}$$

$$= 1 - \frac{6\Sigma d^2}{n(n^2 - 1)} \tag{14.16}$$

Even when the items in question can be given numerical values, the rank correlation coefficient often saves much arithmetic with little loss of accuracy. Thus, in an inquiry whether labour turnover and absenteeism are correlated, turnover and absence rates could be calculated for a number of factories doing similar work, and the factories could be ranked in descending order for both turnover and sickness.

*Example* 14.3
Calculate Spearman's rank correlation coefficient for the data of Example 14.1.

Here $x$ and $y$ denote the ranks of the various towns, beginning with the highest value in each case.

| Town | $x$ | $y$ | $d(=x-y)$ | $d^2$ |
|------|-----|-----|-----------|-------|
| A | 10 | 3 | +7 | 49 |
| B | 3 | 6 | −3 | 9 |
| C | 8 | 1 | +7 | 49 |
| D | 5 | 4 | +1 | 1 |
| E | 2 | 10 | −8 | 64 |
| F | 7 | 2 | +5 | 25 |
| G | 1 | 9 | −8 | 64 |
| H | 9 | 5 | +4 | 16 |
| I | 4 | 7 | −3 | 9 |
| J | 6 | 8 | −2 | 4 |
| Total | | | 0 | 290 |

The sign of $d$ is immaterial, but signs have been inserted here to check that the sum of the differences is zero.

The rank correlation coefficient is, from (14.16),

$$1 - \frac{6 \times 290}{990} = -0.76$$

As usually happens, this agrees fairly well with the value of $r$ found from the original values.

The rank correlation coefficient is often denoted by $\rho$, but as already stated, this symbol is now generally used for the true correlation coefficient of the population from which a sample is taken. Unfortunately, no alternative symbol has been generally accepted.

### Exercises

14.1 What is the object of correlation analysis? Describe and illustrate: (*a*) a scatter diagram; (*b*) a regression line; (*c*) perfect correlation; (*d*) negative correlation.

14.2 Represent the following information regarding to passenger liners on a scatter chart. From an examination of your chart make your best estimate of the speed of a vessel of (*a*) 17,500 gross tons, (*b*) 40,000 gross tons. Discuss the factors affecting the reliability of each of these estimates. Would you feel more confident about one of them than the other; if so, why?

| Gross tonnage | Service speed (knots) | Gross tonnage | Service speed (knots) |
|---|---|---|---|
| 20,900 | 20 | 13,200 | 18 |
| 14,500 | 22 | 21,100 | 22 |
| 6,050 | 16 | 14,300 | 18 |
| 8,800 | 18½ | 28,200 | 24 |
| 13,400 | 15½ | 29,600 | 22 |
| 25,700 | 21½ | 21,300 | 24 |
| 24,200 | 22 | 29,300 | 23 |
| 10,300 | 16 | 24,350 | 23 |
| 7,400 | 16 | 19,400 | 20 |
| 12,450 | 16 | 11,900 | 20 |

(*IoT*)

14.3   An experiment was set up to predict the value of a chemical property of a substance when subject to heat, and the following results were obtained:

| Temperature °C | 20 | 30 | 40 | 50 | 60 | 70 |
|---|---|---|---|---|---|---|
| Values of chemical property | 0·30 | 0·32 | 0·33 | 0·36 | 0·38 | 0·41 |

Find the equation of the line to predict the value of the chemical property given the temperature. (Assume a linear relation.)

What value do you predict for a temperature of 100°C and what assumptions are involved in making this prediction?   (*IoS*)

14.4   Calculate the coefficient of correlation for the following data:

| Expenditure on trading stamps (£) | Sales of petrol (£) |
|---|---|
| 10,000 | 300,000 |
| 25,000 | 500,000 |
| 20,000 | 350,000 |
| 30,000 | 450,000 |
| 15,000 | 400,000 |

Explain the meaning of the term 'coefficient of correlation' and state what its value indicates in this example.   (*ACA*)

14.5   A manufacturer is considering the introduction of an incentive bonus scheme, but fears that it may lead to an increase in the proportion of rejects produced.

The following figures are the results of a sample of shifts worked over the past month

| Production of operative per shift | Percentage of rejects |
|---|---|
| 240 | 1 |
| 450 | 2 |
| 690 | 1 |
| 710 | 3 |
| 730 | 4 |
| 790 | 4 |
| 810 | 5 |
| 890 | 6 |
| 1150 | 6 |
| 1370 | 7 |

(*a*) Calculate a correlation coefficient for these figures.

(*b*) How would you use it to assess the likely results of such a bonus scheme?

(*c*) What reservation would you make in using the coefficient for this purpose? (*ICMA*)

14.6 A destructive test on a range of metal castings has shown a fairly close correlation between measurement $x$ and strength $y$.

| Sample | Measurement $x$ | Strength $y$ |
|---|---|---|
| a | 0·6 | 10 |
| b | 0·9 | 13 |
| c | 1·8 | 18 |
| d | 2·3 | 21 |
| e | 1·7 | 24 |
| f | 2·8 | 26 |
| g | 2·4 | 27 |
| h | 2·8 | 28 |
| i | 3·3 | 31 |
| j | 3·1 | 34 |
| k | 3·8 | 36 |
| l | 4·0 | 40 |

Using these results, compile a formula for estimating strength $y$ from measurement $x$ and predict the strength of castings with the following measurements of $x$: 1·0, 2·2, 2·5. (*ICMA*)

14.7   A company has found that the number of private cars with current licences can be used as a basis for estimating its production requirements. The relevant figures for the past nine years taken from the Annual Abstract of Statistics are as follows:

|      | In thousands |
| --- | --- |
| 1964 | 8,247 |
| 1965 | 8,917 |
| 1966 | 9,513 |
| 1967 | 10,303 |
| 1968 | 10,816 |
| 1969 | 11,228 |
| 1970 | 11,515 |
| 1971 | 12,059 |
| 1972 | 12,717 |

Using the method of least squares, you are required to:

(*a*) estimate the number of private cars with current licences in 1974;

(*b*) provide an estimate for 1974 but assuming that because of other factors, e.g. labour unrest, only 60 per cent of the normal annual increase will be realized in 1974.

Answers should be submitted in millions correct to one decimal place of a million.                                                              (*ICMA*)

14.8   The research department of your company has tested seven types of valve to see how efficiently they performed. The results, together with the prices of the valves, are given below

| Valve | Price (£) | Ranking by performance |
| --- | --- | --- |
| A | 132 | 2 |
| B | 98 | 6 |
| C | 117 | 1 |
| D | 89 | 5 |
| E | 145 | 3 |
| F | 100 | 4 |
| G | 85 | 7 |

Show by calculating a rank correlation coefficient whether in this instance price is a good guide to performance.                         (*ICMA*)

14.9   Ten types of paint have been subjected to artificial 'weathering' tests in the laboratory. The results are listed in the table below, together with those obtained under natural conditions over a longer period of time.

| Type of paint | Ranking obtained: | |
|---|---|---|
| | In the laboratory | Under natural conditions |
| A | 2 | 1 |
| B | 4 | 2 |
| C | 1 | 3 |
| D | 6 | 4 |
| E | 3 | 5 |
| F | 5 | 6 |
| G | 8 | 7 |
| H | 10 | 8 |
| I | 7 | 9 |
| J | 9 | 10 |

Is the laboratory test a reliable guide as to how a given paint will behave under natural conditions?

# 15 Comparing frequencies with expectations

We showed in Chapter 13 how to assess the significance of (*a*) the deviation of a sample mean or proportion from a specified value, or (*b*) the difference between two means or proportions. Tests of proportions are equivalent to tests of the corresponding frequencies, e.g. numbers of defective items in samples. This chapter shows how to extend such tests to more than two categories.

### The chi-squared statistic

Before defining this quantity, it is necessary, in order to explain its name and some of its limitations, to introduce very briefly the statistic correctly called *chi-squared* and denoted by $\chi^2$ ($\chi$ being the Greek letter chi, pronounced 'ky'). If $u_1, u_2, \ldots u_n$ are $n$ random and independent Normal deviates, it is defined by

$$\chi^2 = u_1^2 + u_2^2 + \ldots + u_n^2 \tag{15.1}$$

Clearly there is a separate $\chi^2$-distribution for every value of $n$, which in this case is the number of degrees of freedom. Since, by definition, the expected value of each $u^2$ is 1, the mean of the $\chi^2$-distribution for $n$ degrees of freedom is $n$. The formula for the distribution is complicated and need not be given here. The main point is that these distributions have been tabulated. Table 27 gives, for various degrees of freedom, the values of $\chi^2$ that would be exceeded by chance five times in 100 and once in 100 respectively.

Suppose now that there are $k$ ($>2$) categories or groups into which observations can fall and that the expected frequencies, on some hypothesis which is to be tested, are $e_1, e_2, \ldots e_k$. Then, if the actual

**Table 27** Selected values of $\chi^2$

| Degrees of freedom | $P = 0.05$ | $P = 0.01$ |
|---|---|---|
| 1 | 3·84 | 6·63 |
| 2 | 5·99 | 9·21 |
| 3 | 7·81 | 11·34 |
| 4 | 9·49 | 13·28 |
| 5 | 11·07 | 15·09 |
| 6 | 12·59 | 16·81 |
| 7 | 14·07 | 18·48 |
| 8 | 15·51 | 20·09 |
| 9 | 16·92 | 21·67 |
| 10 | 18·31 | 23·21 |
| 15 | 25·00 | 30·58 |
| 20 | 31·41 | 37·57 |

frequencies are $f_1$, $f_2$, ... $f_k$, the *chi-squared statistic* is defined as

$$\chi'^2 = \frac{(f_1 - e_1)^2}{e_1} + \frac{(f_2 - e_2)^2}{e_2}$$

$$+ \ldots + \frac{(f_k - e_k)^2}{e_k} \tag{15.2}$$

It is so called because, provided no expected number is very small, say less than 5, its distribution approximates closely to the $\chi^2$-distribution for $k - 1$ degrees of freedom. (N.B. $k - 1$, not $k$, because the differences $f_i - e_i$ are not all independent, since $\Sigma f_i = \Sigma e_i$); so closely, in fact, that it is commonly called 'chi-squared' and denoted by $\chi^2$. We shall follow this convention, but the reader should remember that it is only an approximation to the real continuous $\chi^2$.

A high value of $\chi^2$ suggests that the differences between actual and expected frequencies, taken as a whole, are too large to be accounted for purely by chance.

*Example* 15.1

Three factories producing similar goods employ 5000, 3000 and 2000 workers, and in one year experience 80, 66 and 54 accidents respectively. Do these figures indicate real differences between the factories?

The Null Hypothesis is that the risks are equal in all factories, so that in a total of 200 accidents the expected numbers are 100, 60 and 40 respectively. So

$$\chi^2 = \frac{(80-100)^2}{100} + \frac{(66-60)^2}{60} + \frac{(54-40)^2}{40}$$

The calculation can be set out clearly in tabular form:

| Factory | f | e | f − e | (f − e)²/e |
|---------|-----|-----|-------|------------|
| A | 80 | 100 | −20 | 4·00 |
| B | 66 | 60 | + 6 | 0·60 |
| C | 54 | 40 | +14 | 4·90 |
| Total | 200 | 200 | 0 | 9·50 |

Check that the differences $(f - e)$ add up to zero, and be careful with degrees of freedom: 2 here, not 3. The value for $\chi^2$, 9·50, exceeds the 1 per cent value (9·21) for 2 degrees of freedom. It is therefore highly significant, i.e. there are almost certainly real differences in accident risks between the factories.

Notice that $\chi^2$ is only a rough and ready overall test for the observations as a whole. The fact that Factory C makes the biggest contribution to $\chi^2$ does not necessarily mean that it is the odd one out and that there is no real difference between A and B.

## Contingency tables

When observations can be classified in two ways, the data can be set out in a two-way frequency table, called a *contingency table*. If there are, say, 5 rows and 4 columns, there will be 20 *cells* each occupied by a *cell-frequency* (which may be zero). The $\chi^2$ test can be used on such tables to examine whether the two factors are independent or whether there is some kind of association or interaction between them. Example 15.2 has been adapted by kind permission of ICI Ltd from an example given in *Statistical Methods in Research and Production* by Davies and Goldsmith.

*Example* 15.2

The 4 × 3 contingency table given below shows the numbers of times piston-rings have failed in each leg (North, Centre and South) of four compressors over a period of years. Is there any evidence that the proportion of failures in any leg differs significantly in the different compressors?

| Compressor | North Leg | Centre Leg | South Leg | Total |
|---|---|---|---|---|
| A | 22 | 16 | 12 | 50 |
| B | 13 | 10 | 7 | 30 |
| C | 12 | 9 | 19 | 40 |
| D | 16 | 10 | 34 | 60 |
| Total | 63 | 45 | 72 | 180 |

The Null Hypothesis is that the relative risks of failure in the three legs are the same for all compressors, so we calculate the expected frequencies by re-allocating the total for each leg in proportion to the compressor totals. Thus the expected frequency for the cell (C, Centre Leg) is

$$45 \times \frac{40}{180} = 10$$

It may help if we rewrite the table, inserting the expected numbers in brackets. The marginal totals are, of course, the same for both actual and expected frequencies.

| Compressor | North Leg | Centre Leg | South Leg | Total |
|---|---|---|---|---|
| A | 22 (17·5) | 16 (12·5) | 12 (20) | 50 |
| B | 13 (10·5) | 10 (7·5) | 7 (12) | 30 |
| C | 12 (14) | 9 (10) | 19 (16) | 40 |
| D | 16 (21) | 10 (15) | 34 (24) | 60 |
| Total | 63 | 45 | 72 | 180 |

We now calculate

$$\chi^2 = \frac{(22 - 17 \cdot 5)^2}{17 \cdot 5} + \ldots + \frac{(34 - 24)^2}{24}$$
$$= 16 \cdot 83$$

But for how many degrees of freedom? Well, if we fix any three frequencies in the North Leg, the fourth is determined by the column total; similarly in the Centre Leg. But then every number in the South Leg is determined by the row totals, so altogether we have only $3 \times 2 = 6$ degrees of freedom. More generally, in an $m \times n$ table there are $(m - 1)(n - 1)$ degrees of freedom. From Table 27, the 1 per cent value of $\chi^2$ for 6 degrees of freedom is $16 \cdot 81$, just less than the calculated value, so our result is significant at the 1 per cent level.

The $2 \times 2$ contingency table is often used to test the difference between two proportions. The problem posed in Example 13.6 is re-examined in Example 15.3.

*Example* 15.3

Out of 200 articles made on one machine, 22 per cent are faulty; out of 120 made on another machine, 30 per cent are faulty. Is there a significant difference between the machines?

We now deal with actual numbers rather than proportions, and construct the following table.

| Machine | Good | Faulty | Total |
|---------|------|--------|-------|
| A | 156 (150) | 44 (50) | 200 |
| B | 84 (90) | 36 (30) | 120 |
| Total | 240 | 80 | 320 |

The expected numbers are shown in brackets as in Example 15.2; the reader should check them for himself. When one is calculated, the others are fixed by the marginal totals, so there is only one degree of freedom. All differences between actual and expected numbers are equal, viz. 6, although two are positive and two negative.

It is customary with $2 \times 2$ tables to make a continuity correction (cf. Example 12.6) by subtracting $\frac{1}{2}$ from each difference, reducing them here from 6 to 5·5. This improves the approximation to the continuous $\chi^2$ distribution. We now have

$$\chi^2 = \frac{5 \cdot 5^2}{150} + \frac{5 \cdot 5^2}{50} + \frac{5 \cdot 5^2}{90} + \frac{5 \cdot 5^2}{30}$$
$$= 2 \cdot 13$$

This is not significant at the 5 per cent level and, as in Example 13.6, the verdict is 'not proven'. Mathematically, the two methods can be shown to be equivalent.

## Fitting a distribution to frequency data

It is often helpful to know, or at least to be able to assume, that data have come from a particular type of distribution. If, for instance, the parent distribution is Normal, we can use the tables of the Normal distribution. If it is binomial, we are confirmed in our belief that the probability of success is constant; and so on. The usual procedure is to fit a distribution of the specified type to the data, with the same mean and/or SD, and see how good the fit is.

Simple methods, such as drawing histograms or plotting cumulative percentage frequencies on probability paper, are often quite adequate, but it may be desirable, for greater assurance, to perform a *goodness-of-fit test*; this usually involves chi-squared. We shall consider in some detail four of the commonest distributions.

## (a) The rectangular distribution

Strictly speaking, this is a continuous distribution, but the term is often used for discrete distributions with a finite number of possible values, all of which have the same probability.

### Example 15.4

A computer program designed to generate random digits produces 500 digits distributed as follows:

| Digit | 0 | 1 | 2 | 3 | 4 | 5 | 6 | 7 | 8 | 9 |
|---|---|---|---|---|---|---|---|---|---|---|
| Frequency | 40 | 36 | 28 | 62 | 58 | 60 | 34 | 70 | 40 | 72 |

Is there any evidence that the method of generating random digits is unsatisfactory?

The expected frequency for each digit is the same, viz. 50, which simplifies the arithmetic considerably:

$$\chi^2 = \frac{(40-50)^2}{50} + \ldots + \frac{(72-50)^2}{50}$$

$$= \frac{1}{50}(10^2 + 14^2 + \ldots + 22^2)$$

$$= 46 \cdot 56$$

From Table 27, the 1 per cent value of $\chi^2$ for 9 degrees of freedom is $21 \cdot 67$, so the calculated value is highly significant. The program is far from satisfactory—as is, indeed, fairly obvious from the data.

## (b) The binomial distribution

This requires careful thought as to what the parameters $n$ and $p$ represent. The conditions for a binomial distribution are that there should be a definite number $(n)$ of independent items or events, each with two possible outcomes, and a constant probability $(p)$ of 'success' for all events. Unless these conditions obtain, there is no point in trying to fit a binomial. *One should be careful to fit only such distributions as are justified by the nature of the problem.*

*Example* 15.5

Samples of five articles are taken from a process over a long period, and the numbers of defectives in 1024 samples are shown below. Test whether defectives could be occurring randomly with constant probability.

| No. of defectives in sample | No. of samples |
|:---:|:---:|
| 0 | 261 |
| 1 | 369 |
| 2 | 291 |
| 3 | 83 |
| 4 | 20 |
| 5 | 0 |
| Total | 1024 |

As usual in such problems, $n$ is given, viz. 5, and $p$ has to be estimated from the total number of defectives in the 1024 samples, i.e. in 5120 items. There are 1280, giving

$$p = \frac{1280}{5120} = \frac{1}{4}$$

The probabilities of 0, 1, 2, 3, 4, 5 defectives in a random sample are the terms of the binomial expansion of $(\frac{3}{4} + \frac{1}{4})^5$. The expected frequencies are therefore the terms of the expansion of

$$1024(\tfrac{3}{4} + \tfrac{1}{4})^5 = (3 + 1)^5 \qquad \text{(since } 1024 = 4^5)$$

i.e. 243, 405, 270, 90, 15 and 1.

The last two items must be grouped to avoid very small expected numbers. This reduces the degrees of freedom by 1, and another is lost because $p$ had to be estimated from the data. This leaves 3 degrees of freedom.

The calculation is completed as follows:

| No. of defectives in sample | No. of samples Actual | No. of samples Expected | Difference | $\chi^2$ |
|:---:|:---:|:---:|:---:|:---:|
| 0 | 261 | 243 | +18 | 1·33 |
| 1 | 369 | 405 | −36 | 3·20 |
| 2 | 291 | 270 | +21 | 1·63 |
| 3 | 83 | 90 | − 7 | 0·54 |
| 4 or 5 | 20 | 16 | + 4 | 1·00 |
| Total | 1024 | 1024 | 0 | 7·70 |

This is almost significant at the 5 per cent level, leaving a suspicion that the probability of an item being defective is not constant.

### (c) The Poisson distribution

It is always tempting to fit this distribution to one that 'looks like a Poisson', but we must repeat the warning only to fit a distribution for which there are rational grounds. The Poisson distribution applies only to items that can be counted, and is used to test whether events such as arrivals, accidents etc. occur randomly in time or space but at a constant expected rate. This implies that the (small) probability of such an event occurring in the next second, square centimetre etc. is constant. (See Chapter 12.)

One rough test is whether the mean and variance (*not* the SD) are approximately equal. A more exact test is to calculate the mean and fit a Poisson with that value as the parameter (thereby losing 1 degree of freedom as before).

### Example 15.6

Numbers of wireworms in 179 random samples of equal volume of soil were as follows:

| No. of wireworms in sample | No. of samples |
|:---:|:---:|
| 0 | 60 |
| 1 | 74 |
| 2 | 35 |
| 3 | 6 |
| 4 | 3 |
| 5 | 1 |
| Total | 179 |

Test whether wireworms are distributed at random in the soil.

By a strange coincidence (!), the total number of wireworms in these samples is 179, so the mean number per sample is 1. The expected numbers of samples with 0, 1, . . . etc. wireworms are therefore

$$179\, e^{-1}\left[1, 1, \frac{1}{2!}, \frac{1}{3!}, \text{etc.}\right]$$

As before, we set out the calculation in a table, grouping '3 or more' to avoid very small expected numbers.

| No. of wireworms in sample | No. of samples Actual | Expected | Difference | $\chi^2$ |
|:---:|:---:|:---:|:---:|:---:|
| 0 | 60 | 66 | −6 | 0·55 |
| 1 | 74 | 66 | +8 | 0·97 |
| 2 | 35 | 33 | +2 | 0·12 |
| 3 or more | 10 | 14 | −4 | 1·14 |
| Total | 179 | 179 | 0 | 2·78 |

There are now 4 groups only, and the Poisson parameter had to be estimated, so there are 2 degrees of freedom. Table 27 shows that the value of $\chi^2$, 2·78, could easily occur by chance. There is nothing in the data to suggest that wireworms are not randomly distributed in the soil.

### (d) The Normal distribution

One advantage of having Normal data is that one can use the tables of the Normal distribution. On the other hand, a significant departure from Normality may help to pinpoint a fault in a process or a change in operating conditions. These are two good reasons for testing Normality.

### Example 15.7

The following data relate to the percentage ash-content in 280 samples of coal (from BS600, Table 24):

| % ash | No. of samples |
|:---:|:---:|
| 3·0– 3·9 | 1 |
| 4·0– 4·9 | 7 |
| 5·0– 5·9 | 28 |
| 6·0– 6·9 | 78 |
| 7·0– 7·9 | 84 |
| 8·0– 8·9 | 45 |
| 9·0– 9·9 | 28 |
| 10·0–10·9 | 7 |
| 11·0–11·9 | 2 |
| Total | 280 |

It is desired to test whether these sample results could have come from a Normal distribution. We first calculate the mean, 7·35, and the SD, 1·33, using Sheppard's correction (1·36 without it—the effect on the results is negligible). We then convert all group-limits to Normal deviates, remembering that '6·0–6·9', for example, really means 'between 5·95 and 6·95'. Thus, 5·95 becomes:

$$u = \frac{5 \cdot 95 - 7 \cdot 35}{1 \cdot 33} = -1 \cdot 05$$

and 6·95 becomes −0·30.

Now in a Normal distribution, the proportional area $P$ to the left of $u = -1 \cdot 05$ is 0·147, and to the left of $u = -0 \cdot 30$, $P = 0 \cdot 382$, so the area between them ($\Delta P$) is 0·235. With a total of 280, the expected frequency in this range is, to the nearest integer,

$$280 \Delta P = 0 \cdot 235 \times 280 = 66,$$

as compared with the actual frequency 78.

The calculations can be set out as follows:

| Group limit $(x)$ | $u = \dfrac{x - 7 \cdot 35}{1 \cdot 33}$ | $P$ | $\Delta P$ | $E = 280 \Delta P$ | $A$ | $\dfrac{(A - E)^2}{E}$ |
|---|---|---|---|---|---|---|
| 2·95 | −3·31 | 0·000 | | | | |
| 3·95 | −2·56 | 0·005 | 0·036 | 10 | 8 | 0·40 |
| 4·95 | −1·80 | 0·036 | | | | |
| | | | 0·111 | 31 | 28 | 0·29 |
| 5·95 | −1·05 | 0·147 | | | | |
| | | | 0·235 | 66 | 78 | 2·18 |
| 6·95 | −0·30 | 0·382 | | | | |
| | | | 0·292 | 82 | 84 | 0·05 |
| 7·95 | 0·45 | 0·674 | | | | |
| | | | 0·211 | 59 | 45 | 3·32 |
| 8·95 | 1·20 | 0·885 | | | | |
| | | | 0·089 | 25 | 28 | 0·36 |
| 9·95 | 1·95 | 0·974 | | | | |
| 10·95 | 2·71 | 0·997 | 0·026 | 7 | 9 | 0·57 |
| 11·95 | 3·46 | 1·000 | | | | |
| Total | | | 1·000 | 280 | 280 | 7·17 |

The first two groups have been combined, as have also the last two, to avoid very small expected numbers: this reduces the degrees of freedom by 2. Two degrees of freedom are also lost for the parameters that had to be estimated, the mean and the SD. This leaves only 4 degrees of freedom, for which the 10 per cent value of $\chi^2$ is 7·78. Thus there is no strong evidence that the underlying distribution is not Normal, and one cannot go far wrong by assuming that it is.

## General remarks

Like other significance tests, the $\chi^2$-test usually confirms, or fails to confirm, a hypothesis already held for good reasons. It is seldom necessary to calculate $\chi^2$ with great precision; indeed, when the actual and expected frequencies are compared, it may well be unnecessary to calculate it at all.

If $\chi^2$ is not statistically significant, that does not prove the Null Hypothesis. Even a very small value proves nothing either way, although it may suggest that someone is cooking the figures. Agreement between actual and expected frequencies can sometimes be too good!

## Exercises

15.1   The table below gives details of the numbers of accidents occurring in different areas of a factory last year. Is there a significant difference between the chances of a worker being involved in an accident in the different areas?

|   | Area | No. of Accidents | Average No. of Workers Employed |
|---|---|---|---|
| 1 | Production | 169 | 700 |
| 2 | Packaging | 32 | 100 |
| 3 | Quality Control | 3 | 20 |
| 4 | Engineering | 15 | 40 |
| 5 | Laboratory | 16 | 60 |
| 6 | Warehouse | 15 | 80 |

*(IoS)*

15.2   Bricks made in two kilns have been graded as facings (high quality), second (medium quality) and commons (rather poor quality). The production of bricks in a particular period was as follows:

|  | Facings (000's) | Seconds (000's) | Commons (000's) | Total (000's) |
|---|---|---|---|---|
| Kiln A | 24 | 43 | 13 | 80 |
| Kiln B | 31 | 57 | 32 | 120 |

It is the policy of the business to produce bricks of the highest quality possible and the number of seconds and commons produced is due to the efficiency or inefficiency of the process. Use the $\chi^2$ test to determine whether the manager of kiln A is justified in claiming that he produces bricks of a higher quality than his colleague at kiln B.

*(ICMA)*

15.3   A company which manufactures Spiv washing machines conducted a survey of 500 women who bought their machines last year. Only 150 of these 500 women replied to the postal questionnaire, so the company decided to send reminders to the remaining 350 women. This produced a further 200 replies and a second reminder produced another 50 replies. The remaining 100 women were visited by an interviewer and they all gave the required information at this stage. As a result of this work, the company suspects that there is some association between a woman's degree of satisfaction with her Spiv washing machine and her willingness to complete the questionnaire. Do the figures given below support this hypothesis?

| | Replied immediately | Replied to first reminder | Replied to second reminder | Visited by interviewer | Total |
|---|---|---|---|---|---|
| Satisfied | 100 | 134 | 21 | 45 | 300 |
| Indifferent | 40 | 48 | 20 | 42 | 150 |
| Not Satisfied | 10 | 18 | 9 | 13 | 50 |
| *Total* | 150 | 200 | 50 | 100 | 500 |

*(IoS)*

15.4   The following data relating to abortions in England and Wales during the year 1970 is for the age group 25–29, for cases not resident in England and Wales.

| Marital Status | Social Class | | | |
|---|---|---|---|---|
| | I | II | III | IV & V |
| Single | 34 | 287 | 519 | 88 |
| Married | 63 | 170 | 380 | 64 |
| Other | 4 | 30 | 83 | 18 |

Is there evidence of an association between marital status and social class?                                                                      *(IoS)*

15.5   A new treatment to prevent deep vein thrombosis (DVT) after major surgery was randomly allocated to incoming patients in one hospital. After several months the investigating surgeon drew up the following table:

|             | DVT | No DVT | Total |
| ----------- | --- | ------ | ----- |
| Treated     | 7   | 30     | 37    |
| Not treated | 16  | 18     | 34    |
| Total       | 23  | 48     | 71    |

Do these data indicate that the treatment was effective?

(IoS)

15.6   In a study of the distribution of tree species in a region of tropical rain forest, 86 square areas (quadrats) were surveyed, and all trees with trunk girth greater than 20 cm were recorded. For one species the distribution of specimens per quadrat was as follows:

| Number of specimens | 0  | 1  | 2 | 3 | 4 |
| ------------------- | -- | -- | - | - | - |
| Number of quadrats  | 48 | 25 | 6 | 4 | 3 |

What distribution would you expect these data to fit on the assumption that trees of the species are randomly distributed throughout the forest? Fit the distribution, and test the assumption of randomness.

(IoS)

# 16 Some industrial applications of statistics

This chapter may almost be regarded as a continuation of Chapter 13, as it deals principally with certain applications of sampling theory to industrial problems. They are made necessary by the fact of variability in processes and experiments.

### Variation in industrial processes

No matter how good the equipment or how skilful the operative, a number of articles produced by the same process will never be exactly alike, although with imperfect measuring instruments they may appear so. There will always be that slight variation from one to another, as if the material were endowed with an element of free will.

The causes of variation, which are many, may be divided into *assignable causes*, such as mechanical faults or abnormalities in the raw material or in the reaction conditions of a chemical process, which can be identified and eliminated, and *chance causes* which are either unknown or impossible to eliminate. When only chance causes are operating, the process is said to be *under control*. The data recorded will then vary about the *process average*, generally denoted by $\bar{X}$, with a standard deviation $\sigma$ measuring the *residual variation*, i.e. the inherent variability of the process.

### Control charts for variables

A useful device for helping to maintain the efficiency of a process is the *control chart*, on which observations are plotted, generally at regular intervals of time. The use of control charts for this purpose is called *Quality Control* or, to give it its full title, *Statistical Quality Control*.

We shall assume for simplicity that there is only one variable of interest to us, that residual variations are Normally distributed and, unless there is evidence to the contrary, that the process is under control. To ensure that it remains so, we must keep a check on both the mean and the variability. To do this, we take small samples, usually about 5 and rarely more than 10, and record their means and ranges on separate charts (but preferably on the same sheet).

At first sight, the use of the range is surprising since, as we remarked in Chapter 11, a single range is an unreliable measure of dispersion (but so is the SD of a very small sample), and moreover the range tends to increase with the size of sample. However, it is much easier for the operative (or even the supervisor) to understand and to calculate than the standard deviation. Furthermore, the mean of 20 or more sample ranges will provide a good estimate of the SD if required. Suppose, for example, the mean range of 20 or more samples of 5 items is 7 cm. The expected range of 5 random observations from a Normal distribution is $2 \cdot 33\sigma$, so we take $\sigma$ to be

$$\frac{7}{2 \cdot 33} = 3 \text{ cm almost exactly}$$

The chart for means contains horizontal lines representing (i) the process average, and (ii) *control limits* outside which only a small specified proportion of sample means will fall when the process is under control (see Fig. 44). It is customary to show two pairs of control limits, the *inner control limits* or *warning limits* containing all but 5 per cent of such observations and the *outer control limits* or *action limits* all but 0·2 per cent. As with levels of significance, experience shows that these percentages are convenient as a rough guide.

If $\sigma$ is known, the control limits for the mean are easily calculated. The standard error of the mean of $n$ items is $\sigma/\sqrt{n}$, so the warning limits, which contain 95 per cent of all such means, are

$$\bar{X} \pm 1 \cdot 96 \frac{\sigma}{\sqrt{n}}$$

and similarly the action limits are

$$\bar{X} \pm 3 \cdot 09 \frac{\sigma}{\sqrt{n}}$$

These limits are commonly expressed in terms of the mean range $\bar{w}_n$ instead of $\sigma$. (The suffix can be dropped if there is no ambiguity.) Thus, the action limits can be expressed as

$$\bar{X} \pm A_{0 \cdot 001} \, \bar{w}_n$$

**Fig. 44** Control chart for the mean length of a sample of five rods

where

$$A_{0.001} = \frac{3.09}{\sqrt{n}} \frac{\sigma}{\bar{w}_n}$$

For example, if $n = 4$, $\bar{w}_4 = 2.06\sigma$, so

$$A_{0.001} = \frac{3.09}{2.06\sqrt{4}} = 0.75$$

The factors $A_{0.025}$ and $A_{0.001}$ have been tabulated for various values of $n$.

The range chart shows the mean range and control limits (see Fig. 45). The distribution of the sample range $w_n$ is distinctly skew, especially for very small values of $n$, so the control limits are not symmetrical about the mean range. Factors $D_{0.025}$ etc. have been calculated by which $\bar{w}_n$ should be multiplied to find the control limits. Thus, for $n = 5$ the warning limits are $0.37 \bar{w}_5$ and $1.81 \bar{w}_5$ and the action limits are $0.16 \bar{w}_5$ and $2.34 \bar{w}_5$. The lower control limits are often omitted, because reduced variation is a cause for satisfaction, not for concern.

**Fig. 45**   Control chart for the range of the lengths of a sample of five rods

*Example* 16.1

A machine is turning out steel rods with a process average of 6 cm and a standard deviation of 0·005 cm. Construct control charts for the mean and range of the lengths of samples of five rods.

The standard error of the mean of five items is

$$\frac{0·005}{\sqrt 5} = 0·00224 \text{ cm.}$$

Hence the inner control limits are

$$6 \pm 1·96 \times 0·00224 = 6 \pm 0·0044 \text{ cm}$$

and the outer control limits are

$$6 \pm 3·09 \times 0·00224 = 6 \pm 0·0069 \text{ cm.}$$

The control chart for the mean is therefore as shown in Fig. 44.

For the range, $\bar w = 2·33 \times 0·005 = 0·0116$ cm, the inner control limits are

$$0·37 \times 0·0116 = 0·0043 \text{ and } 1·81 \times 0·0116 = 0·0210 \text{ cm,}$$

and the outer control limits are

$$0·16 \times 0·0116 = 0·0019 \text{ and } 2·34 \times 0·0116 = 0·0272 \text{ cm.}$$

Fig. 45 shows the resulting control chart.

The means chart shows that some action was taken following an observation outside the action limits. It is fairly certain that the mean has 'drifted', but there is still a possibility that the observation was a random though unusual one. The risk of making a wrong decision has to be taken—either way, for there is also the risk of *not* noticing for some time that the mean has drifted. If there are specification limits, e.g. if rods longer than 6·02 cm or shorter than 5·98 cm are unacceptable, this may mean a certain proportion of waste product. The two kinds of risk are inherent in all statistical sampling.

The control limits are only a guide, not a substitute for common sense. For example, two or three consecutive points just inside the inner control limits would suggest that the process needed investigating. Whether it were actually stopped for inspection would depend on the nature of the process, the cost of interrupting it, and so on.

## Control charts for attributes

The control charts described above are suitable for substances like coal, glass and fibres which can be measured in terms of one or more continuous variables such as density, length, tensile strength or the percentage of some ingredient; but there are manufactured articles like lamps, valves and sacks that cannot be measured in this way but are either good or bad. Such things can be checked only by taking fairly large samples and finding the proportion defective.

Suppose, for example, that a machine is producing large quantities of small articles of which, under normal conditions, 3 per cent are defective, and that samples of fifty are tested from time to time. Then the average number of defectives in a sample will be $1\frac{1}{2}$ and the actual number of defectives will be distributed in a Poisson distribution with $m = 1·5$. The numbers observed in samples of fifty can be plotted on a control chart as in Fig. 46.

One difficulty here is that no control limit can be fixed such that exactly 5 per cent of the observations are above it and 95 per cent below. In this example nearly 2 per cent of the samples will contain five or more defectives, and about $6\frac{1}{2}$ per cent will contain four or more. Accordingly, a control line has been drawn between four and five, and the probability of a point lying above that line by chance is about 0·02.

If the 5 per cent confidence level is particularly desired, it can be obtained by finding a suitable sample size.

**Fig. 46** Control chart for numbers of defectives in samples of fifty with a process average of 3 per cent defective

## Cumulative sum charts

A useful alternative to the type of chart described above is the *cumulative sum chart*, in which the item plotted is not the number of defectives, say, in each individual sample, but the total number of defectives found in all samples taken to date, starting at a given time and continuing for a long period. Fig. 47 shows such a chart for the same data as were depicted in Fig. 46.

Over a fairly long period, the slope of the chart will provide an estimate of the average quality of the product, and a marked change in the slope will indicate a change in quality. These charts can also be used with continuous variables, the quantity cumulated being not the actual length etc., but the deviation of that variable from the process average or standard value.

## Advantages of control charts

The great advantage of control charts is that they enable a continuous watch to be kept on the process, thereby saving wastage of product. Furthermore, they provide a permanent record of performance to which customers can be referred to satisfy them that the product is of good quality. They also have a psychological effect in keeping operatives alert for faults and giving them a greater interest in their work. Control charts can be applied to many other fields besides industrial processes, e.g. accidents, typing errors, sickness rates and so on.

**Fig. 47**   Cumulative sum chart for data of Fig. 46

## Acceptance sampling

*Acceptance sampling* generally means the use of sampling methods to check the quality of purchased materials, but it can be applied to auditing of accounts and invoices and other forms of checking. It may be a single test for a particular *batch* or *lot*, or it may be a series of checks of successive lots, in which case the control-chart technique can be used.

The difficulty with single lots is that unless the sample is very large it may give quite a misleading result. The tester must therefore take a certain risk (*consumer's risk*) of accepting an unsatisfactory batch, e.g. one containing 5 per cent or more defectives, and also a risk (*producer's risk*) of rejecting a batch that is actually good, e.g. containing less than 1 per cent defective. The actual quantities chosen are largely a matter of judgment.

Suppose, for example, that a purchaser wishes to be 95 per cent confident that a particular lot does not contain more than 2 per cent defective. Clearly, to give this degree of confidence, the percentage defective in the sample must be considerably less than 2 per cent. Thus, a sample of 150 from a lot with 2 per cent defective would contain only one defective about once in five times, so it is easy for a fairly large sample containing less than 1 per cent defective to mask a proportion of 2 per cent defective in the lot.

Consider the following plan:

(*a*) Take a sample of 150 items and, if no defectives are found, accept the lot. Otherwise, reject it.

This plan would ensure that 95 per cent of all lots with 2 per cent defective were rejected, but it would also result in the rejection of many very good lots simply because the rare defective happened to occur in the sample. For example, it would reject 53 per cent of all lots containing only $\frac{1}{2}$ per cent defective. Clearly, a plan that allows no defectives whatever is not sufficiently discriminating.

Now consider a slightly more selective plan:

(*b*) Take a sample of 240 items, and accept the lot if there is not more than one defective. If there are two or more, reject it.

This plan would also reject 95 per cent of all lots containing 2 per cent defective, but it would reject only 34 per cent of those containing $\frac{1}{2}$ per cent defective. For a really selective plan it would be necessary to take a much larger sample.

Plans (*a*) and (*b*) are examples of *single sampling*, but *double sampling*, or sampling in two stages, is often more efficient and economical. The essence of double sampling is that if the first sample is indecisive, the result is decided on a second. On the average, this reduces the total amount of sampling, particularly where the quality of a lot is likely to be either very good or very bad.

The principle of double sampling can be extended to *sequential sampling*, in which observations are taken one at a time until a decision can be reached on the accumulated data, no sample size being fixed beforehand. After every observation a decision is taken either to accept the batch, or to reject it, or to continue sampling. Sequential sampling requires, on the average, much smaller samples than equivalent single or double sampling schemes, but the theory is complicated and sequential schemes are more difficult for an unskilled person to operate.

The words 'accept' and 'reject' have been used in a general sense, but rejection does not necessarily mean that the product is wasted. It may mean inspecting every item and replacing faulty ones. This is called *rectifying inspection*. Alternatively, the product may be sold at a reduced price.

Acceptance sampling usually requires large samples to give a reasonable degree of confidence in the quality of individual batches, and it is generally more economic and more satisfactory to keep a general check on quality by means of control charts, or to know that the supplier is doing so. In all sampling schemes it is necessary to balance the cost of inspection against the cost of occasional wrong decisions or failure to detect defective products. The ideal, in fact, is to minimize the average total cost in the long run.

## Operational research

Operational research is the application of scientific method, including statistical method, to the various operations of industry, commerce, warfare etc. In industry, this covers not only the manufacturing processes but such matters as stock control and distribution. In fact, these problems cannot be considered most efficiently in isolation. Stock policy, for example, will react on production planning, and both will depend on good sales forecasting. It will be convenient, however, to begin by considering the problem of optimum stocks.

Since demand is generally irregular and, to a large extent, unpredictable, it is often necessary to keep large stocks of finished goods in order to be able to meet any reasonable demand. If stocks are too low they will run out frequently and the manufacturer will lose business. On the other hand, if stocks are so high that it will hardly ever be necessary to refuse or delay an order, the cost of holding them may be prohibitive. Besides, some products may deteriorate or become obsolete if kept too long. It is therefore necessary to find the best compromise, and to balance the cost of holding stocks against the cost of occasionally being unable to fulfil an order promptly. Similarly with stocks of raw materials, it is necessary to balance the cost of storage against the cost of occasionally running out of materials and holding up production, or having to take emergency measures to obtain materials quickly.

Very often the problem can be solved by a *simulation study* based on past experience, on the assumption that the pattern of past demand can be taken as a fair guide to the future. The procedure is to take several years' data of, say, weekly or monthly demands, to construct a large number of possible sequences of orders by random sampling from those data and, assuming various levels of safety stock, to ascertain how often stocks would run out and what the total cost would be in each case, and then choose the safety stock for which the total cost is least. This technique, popularly called 'Monte Carlo methods', can be used very effectively with the aid of an electronic computer.

The same technique can be employed with production planning. For example, in a firm making a wide range of products, e.g. varieties of paint, it may be a constant problem what to make next, and in what quantity. Probably the plant has to be cleaned out after one batch has been made before the next batch, of a different variety, can be started. It would be uneconomic either to make very small batches or to make such large batches as to create excessive stocks. Again the best compromise has to be found, and this problem, together with that of optimum stocks, can be programmed for a computer.

A rather similar problem is that of the best distribution of a product from a number of factories or depots to a large number of customers, given various transport costs, the total demand of each customer, the total capacity of each factory etc. There are many other problems occurring in operational research, such as the right number of machines for one operative to tend when the machines break down at irregular intervals, and the best form of telephone or teleprinter network connecting the works and offices of a large company. The development of electronic computers has greatly facilitated the solution of such problems.

## Statistical methods in industrial research

In the research laboratory there is scope for statistical methods to deal with the uncertainty caused by experimental errors, instrumental errors, variations in materials etc. Frequently, in an investigation to compare two methods or materials, say, or to estimate some physical quantity, one experiment is not enough. Too many would be wasteful, but too few would mean too great a risk of drawing wrong conclusions. Statistical methods help to determine the right number of experiments to make, and to assess the precision of the result.

It often happens that the yield or quality of a product depends on several factors, like temperature, reaction time and strength of solution. If these factors were independent of one another, it would be possible to determine first the optimum temperature, then the optimum reaction time, and so on. In practice, the problem is often complicated by interactions between the various factors. For example, the best temperature or pressure for one strength or reaction time may not be the best for another. Frequently the factors are so numerous, and can be varied in so many ways, that it is impossible or uneconomic to test every possible combination of them. It is necessary then to design a limited number of experiments in such a way as to extract the maximum amount of information at the minimum cost in time and money. Such *factorial experiments* can be designed, and the results analysed, by the aid of statistical methods.

This chapter has only touched the fringe of a vast subject which is still developing rapidly. For further information the reader should refer to more specialized books on industrial statistics, quality control, design of experiments etc.; but perhaps enough has been said to show that statistics is neither a tool for the amateur nor a toy of the classroom, but a subject of great practical importance to industry as well as to the natural and social sciences.

## Suggestions for further reading

Davies, O. L. & Goldsmith, P. L. (eds). *Statistical Methods in Research and Production*. London: Oliver and Boyd.

Davies, O. L. (ed.). *Design and Analysis of Industrial Experiments*. London: Oliver and Boyd.

Duckworth, W. E. *A Guide to Operational Research*. London: Methuen.

Finney, D. J. *An Introduction to the Theory of Experimental Design*. University of Chicago Press.

Huitson, A. & Keen, J. *Essentials of Quality Control*. London: Heinemann.

Makower, M. S. & Williamson, E. *Teach Yourself Operational Research*. English Universities Press.

Paradine, C. G. & Rivett, B. H. P. *Statistics for Technologists*. English Universities Press.

Wetherill, G. B. *Sampling Inspection and Quality Control*. London: Methuen.

## Exercises

16.1   Reels of insulated wire are made such that the lengths are Normally distributed with a mean length of 1000 yd and a standard deviation of 10 yd. Two of these reels are selected at random and wound together to make a twin flex package. When one of the reels contributing to this package runs out, the wire remaining on the other package is regarded as waste. What average waste may be expected?

It is decided that a check will be kept on the amount of waste by actually measuring the waste from packages sampled at fairly frequent intervals. Set up a control chart and design record forms to do this.

(*IoS*)

16.2 The following are weights of medicinal tablets which are nominally 10 grains. Ten samples of 5 tablets have been taken at hourly intervals from the manufacturing process.

| | | | | | | | | | |
|---|---|---|---|---|---|---|---|---|---|
| 10·1 | 10·0 | 10·1 | 9·8 | 9·9 | 10·0 | 10·0 | 9·7 | 9·9 | 10·0 |
| 10·0 | 9·8 | 9·9 | 10·2 | 9·9 | 10·0 | 10·1 | 9·9 | 9·8 | 10·2 |
| 10·0 | 9·9 | 10·1 | 10·1 | 10·0 | 10·0 | 9·4 | 10·2 | 10·0 | 10·0 |
| 10·1 | 10·2 | 9·9 | 9·9 | 10·0 | 10·1 | 10·0 | 10·0 | 10·1 | 9·7 |
| 10·3 | 10·0 | 10·1 | 9·9 | 10·0 | 10·1 | 10·2 | 9·8 | 9·8 | 10·0 |

(*a*) Construct and plot suitable Quality Control Charts for this process.

(*b*) State and explain five criteria for detecting any changes which might occur in the manufacturing process. (*IoS*)

16.3 The mean weight of a large loaf produced at a bakery was found in a series of quality control checks to be 30 ounces, with a standard deviation of 1 ounce.

Sample batches of 4 loaves were then taken from the travelling oven at 2-hourly intervals and the following are the averages of the last 6 consecutive sample batches:

29·6, 29·4, 29·9, 30·2, 30·3, 30·1

(*a*) Draw up a control chart for these averages and enter the figures on the chart.

(*b*) What proportion of loaves would you expect to fall below the legal minimum weight of 28 ounces? (*ICMA*)

16.4 A machine makes bolts in batches of 1024. Twenty consecutive batches were tested and the number of defective bolts in each is given below:

(i) Draw a control chart for the process showing control limits at any two levels you consider significant.

(ii) Enter the results shown below on to your chart and comment on the results. (*IoS*)

| Batch | Number of defective bolts | Batch | Number of defective bolts |
|-------|---------------------------|-------|---------------------------|
| 1 | 2 | 11 | 2 |
| 2 | 3 | 12 | 1 |
| 3 | 7 | 13 | 5 |
| 4 | 1 | 14 | 4 |
| 5 | 0 | 15 | 3 |
| 6 | 4 | 16 | 12 |
| 7 | 8 | 17 | 8 |
| 8 | 7 | 18 | 1 |
| 9 | 1 | 19 | 2 |
| 10 | 3 | 20 | 2 |

(*IoS*)

16.5 For a batch process making ceramic parts in large batches of approximately equal size, the percentage wastage is estimated by weighing the defective parts and quoting this weight as a percentage of the average weight of a batch. The 20 percentages tabled below are the percentage wastage in each of twenty successive batches.

| Batch number | 1 | 2 | 3 | 4 | 5 | 6 | 7 | 8 | 9 | 10 |
|--------------|---|---|---|---|---|---|---|---|---|----|
| Per cent wastage | 5·2 | 4·2 | 2·4 | 6·4 | 6·0 | 5·2 | 2·8 | 7·8 | 4·6 | 2·8 |

| Batch number | 11 | 12 | 13 | 14 | 15 | 16 | 17 | 18 | 19 | 20 |
|--------------|----|----|----|----|----|----|----|----|----|----|
| Per cent wastage | 3·4 | 5·3 | 2·9 | 4·6 | 2·5 | 6·5 | 4·1 | 7·3 | 3·5 | 7·5 |

Construct a control chart including and employing these first 20 observations, and discuss whether the process is in control. (*IoS*)

16.6 The following figures were obtained from a series of quality control checks on engineering components, six items being measured at hourly intervals.

| Time | 9 am | 10 am | 11 am | 12 noon | 1 pm | 2 pm |
|------|------|-------|-------|---------|------|------|
| Sizes | 119·0 | 120·8 | 119·7 | 119·9 | 118·5 | 119·7 |
| | 120·9 | 119·8 | 120·1 | 119·2 | 119·6 | 118·9 |
| | 119·0 | 118·9 | 120·9 | 121·1 | 120·2 | 123·1 |
| | 121·3 | 119·2 | 118·0 | 120·3 | 122·3 | 121·2 |
| | 120·0 | 120·5 | 118·6 | 122·6 | 122·9 | 118·0 |
| | 121·0 | 120·2 | 120·3 | 119·3 | 120·1 | 119·7 |

Depict these data on mean and range charts, using the following limits:

For the mean chart: Inner limits 119·5 and 120·5

Outer limits 119·2 and 120·8

For the range chart: Inner limit 3·5

Outer limit 4·5

Comment on the results. (*ICMA*)

16.7 A certain assembly operation must be performed in a temperature- and humidity-controlled, dust-free atmosphere. To check this, an air sample is taken at hourly intervals and the dust particles above the critical size in a fixed volume of air are counted. The *twenty-five* counts taken during the first complete day, midnight to midnight, of the air-sampling system were: 0, 1, 1, 2, 3; 0, 1, 0, 0, 2; 2, 1, 1, 2, 0; 0, 3, 4, 0, 0; 0, 1, 0, 0, 1.

Construct an attribute control chart for the air-sampling and state the assumptions used in constructing the chart. Do these samples indicate a stable level for the dust content of the air? (*IoS*)

16.8 The following numbers of defectives were obtained in samples of 50, taken at regular intervals:

| Sample number | Number of defectives | Sample number | Number of defectives |
|---|---|---|---|
| 1 | 10 | 11 | 8 |
| 2 | 5 | 12 | 14 |
| 3 | 19 | 13 | 5 |
| 4 | 16 | 14 | 12 |
| 5 | 14 | 15 | 17 |
| 6 | 18 | 16 | 10 |
| 7 | 7 | 17 | 14 |
| 8 | 2 | 18 | 17 |
| 9 | 11 | 19 | 21 |
| 10 | 8 | 20 | 24 |

Construct a control chart and comment as necessary. (*IoS*)

16.9 Batches of 100 components were taken at fixed intervals of time from a production line and tested. The following figures are the number of components found to be defective in each of the batches:

4 3 2 0 7 5
2 4 1 3 4 6

(*a*) Calculate the mean proportion defective.

(*b*) Draw up a quality control chart for samples of 200 of the components, such that the inner limit is equal to the mean plus one standard error and the outer limit is equal to the mean plus two standard errors.

(*c*) Could your chart be used for batches of 50 components? If not, why not? (*ICMA*)

16.10  A process has a target mean of 40 g with a known SD of 4 g. Samples of size 4 are drawn and the results plotted on Shewhart Mean & Range Charts.

(*a*) What are the appropriate confidence limits for these charts?

(*b*) Comment on the practical implications of the following results which appear on the Shewhart Mean Chart:

    (i) A reading of 42·4 g.

    (ii) A reading of 35·8 g.

    (iii) Two successive readings of 44·7 and 43·8 g.

    (iv) Two successive readings of 44·5 and 35·7 g.     (*IoS*)

# 17 Index numbers

Most readers will be familiar with the idea of an *index number* as a device for estimating trends in prices, wages, production and other economic variables. There is nothing abstruse about an index number. In its simplest form it represents a special case of an average, generally a weighted average, compiled from a sample of items judged to be representative of the whole.

Suppose an index of commodity prices is required. There are many commodities, with prices expressed in different units, but there is usually a tendency for prices of allied commodities to rise or fall together. Imagine the price of each commodity plotted on a graph in the form of daily or weekly quotations, or monthly or annual averages. Two or more such prices can be represented on the same graph if they can be reduced to the same units or dimensions, and a standard way of doing this is to express the current price as a percentage of the corresponding price at some fixed date in the past, called the *base date*, or of the average price for some given period, the *base period*. This percentage, known as a *price relative*, is independent of the units in which the price is quoted. The resulting graph will show a set of irregular lines fanning out from a single point corresponding to the base date or period, when each price relative is equal to 100. Unless these lines are very erratic, however, it should be possible to detect a general trend in the graph as a whole, corresponding to the movement of prices in general which it is the function of the index number to measure.

If the index number is obtained as an average of price relatives (as will be shown below, that is not the only way), it follows that the value of the index for the base date or period is also 100. In algebraic formulae, however, it will be convenient to omit the factor 100. The weights (if any) allotted to the various items will depend on their relative importance, and since the importance of a commodity varies with time, place and the means and habits of those who use or trade in it, it is clear that there is no single solution to the problem, in the sense that one solution is right and all others are wrong. Fortunately, however, at least over short periods, solutions reached by different methods and with different weighting systems seldom differ by more than a few points.

The weights may either be fixed or vary from year to year, or even month to month. When fixed weights are used, they often represent sales or similar quantities relating to a particular date or period, called the *weights base*. If, as generally happens, the weights base coincides with the *date base* (the base date or period defined above), the terms 'base date' and 'base period' may be used without ambiguity. If the weights base and date base differ, the terms 'base date' and 'base period' are taken to refer to the date base. If the weights are arbitrary or variable there is, of course, no weights base.

Unless otherwise stated it will be assumed that the weights base (when it exists) is also the date base. It will also be convenient to take the base period as a year and to refer to index numbers of prices. In general, however, the principles and methods described in this chapter apply equally to indices of sales, production, wage-rates etc., and to other base periods.

## Choice of items

The first, and probably the biggest, problem is to decide what items should be included in the index. They should be (1) relevant, (2) representative, (3) reliable, and (4) comparable over a period of time.

1. Items must be relevant, e.g. a producers' price index should be composed of wholesale prices, and a cost of living index should comprise retail prices together with rent, rates etc. One should not use an index of retail prices to estimate changes in construction costs, or an index of wage rates of manual workers to estimate changes in the level of salaries of business executives.

2. An index number is generally based on a sample. For example, it would be impossible to include in an index of food prices every conceivable variety of food. It is desirable, therefore, that the items chosen should be adequate in number and importance, and representative of the whole. Generally, the more items there are, the more reliable will the sample be, and the less likely to be unduly affected by abnormal movements in one or two commodities. After a certain point, however,

greater precision is outweighed by the expense and labour of collecting additional information.

In some cases, e.g. where a price index is being compiled for the products of a firm or of a fairly simple industry, it may be possible to cover the bulk of them with relatively few items. Often, however, a commodity is sold in many different sizes, grades, packages etc., and under various contracts, and it may be necessary to select some specified terms of sale as giving a representative price.

3. As has been said before (it is worth repeating), it is essential for the basic data to be as accurate as possible. Preference should be given to items for which reliable quotations can be obtained with confidence. No theoretical refinements in formulae or methods of computation can compensate for serious deficiencies in the data.

4. One of the greatest difficulties is variation in quality. Very often it is only possible to obtain an average price for a commodity like paint, which may vary tremendously in composition and quality. Similarly with iron and steel and other metals, where a change of average price may merely reflect changes in the type, grade etc. of the product. In such cases we have to be content with a unit value index, derived not from prices for specified grades, contract terms etc., but from average values per ton or other unit of what may be very mixed bags of products. One simply hopes that changes of internal pattern average out fairly well.

It may also be necessary to exclude items where classification has changed or because data are not available for both base year and current year. A special case of this is products that were not known or not fully developed in the base year. Sometimes a new product is given a fictitious base-year price, but this is not very satisfactory. Another way of introducing new products is the chain base method (see page 222).

## Sources of data

When the items composing the 'basket' have been decided, the next question is where the figures are coming from. Sometimes price quotations can be obtained from trade journals or from leading firms of dealers in the commodities in question, or they may be based on average import or export prices. Failing that, it may be necessary to take representative prices as reported by local observers (see Chap. 20).

Production data may be obtained similarly from published sources, trade associations or (by the Government) from producing firms.

## The base period

The base period chosen is generally a year, and a fairly 'normal' year for preference. If there is any difficulty in selecting a normal year, the average of a series of years can be taken. Thus, *The Statist* Wholesale Price Index Numbers were for many years based on the period 1867 to 1877, and several American indices are based on a period of three years.

For indices of production, sales or other quantities, it is nearly always necessary to take at least a year as base period, since any shorter period might be unrepresentative and reflect seasonal movements; variation in the length of the months must also be allowed for. For many prices, wages rates etc., it may be sufficient to take a particular day or the average for a single month. For example, the Board of Trade Index Numbers of Wholesale Prices were at one time based on 30 June 1949.

## Progressive changes in index numbers

For various reasons, most index numbers are subject to revision from time to time. Major changes may be due to far-reaching changes in conditions, habits and standards of life reflected in the pattern of expenditure, or to new ideas and techniques resulting in the adoption of new formulae. Minor changes may be caused by the inclusion of new items or the disappearance of old ones, or by periodical revision of the weighting system.

When a major change is made it is customary to begin again with a new base, frequently the year on which the new weighting system is based. In this case it is not usually necessary to recalculate the old index number on the same system; it is sufficient to splice or blend the two series as follows. Suppose the old series is:

1969—100
1970—108
1971—111
1972—120
1973—125

and the new series is:

1973—100
1974—114
1975—121
1976—124

The whole of the first series is multiplied by 0·8, the reciprocal of 1·25, so that the index for 1973 is 100 in both series. The combined series is then:

1969— 80
1970— 86
1971— 89
1972— 96
1973—100
1974—114
1975—121
1976—124

## Methods of calculation

Broadly speaking, methods of calculating an index fall into two categories, one based on *relatives* and the other on *aggregates*. In certain circumstances, as will be shown later, the two methods are equivalent.

A price relative is simply the current price of a commodity divided by the base price. Like the index itself, it is generally expressed as a percentage. Thus, if the price of an article is £15 and its average price in the base year was £12, the price relative will be:

$$\frac{15}{12} \times 100 = 125$$

Relatives of output or other quantities are defined similarly. An obvious method of compiling an index, therefore, is to calculate relatives for all constituent items and average them in some way. The next few sections will describe several ways of doing this.

The other method is to consider an aggregate or 'basket' of goods, e.g. the goods actually bought or sold in a given year, and compare their value at current prices with their value at base-year prices. This method will also be discussed in some detail.

Naturally, these different methods will give slightly different results. This is not a serious objection to index numbers as such, for unless price movements are very erratic or conditions are changing so greatly that comparisons mean little in any case, it will be found that the differences are not serious. Some methods are better and some indices more accurate than others, but it is seldom possible to say that one is right and the others wrong. An index number is like a slide rule; its purpose is to give a fair estimate, and this it generally does.

## The arithmetic mean of relatives

One of the simplest ways of constructing an index number is to take the arithmetic mean of the relatives of the various items, as shown in Example 17.1.

*Example* 17.1

Calculate a price index for 1976 based on 1975 as 100 for a group of three commodities whose average prices in 1975 and 1976 were as follows:

| Commodity | 1975 Price | 1976 Price |
|---|---|---|
| A | £40 per ton | £50 per ton |
| B | £ 5 per cwt | £ 7 per cwt |
| C | £50 per gross | £58 per gross |

It is easily seen that the price relatives for $A$, $B$ and $C$ are 125, 140 and 116 respectively. Their simple arithmetic mean is:

$$\tfrac{1}{3}(125 + 140 + 116) = 127$$

In a simple AM it is implied that all items are equally important, but if (as generally happens) they are not, the relatives can be weighted by appropriate factors. Suppose, therefore, that it had been decided to give Commodities $A$, $B$ and $C$ weights of 6, 3 and 1 respectively; then the index for 1976 (1975 = 100) would have been:

$$\frac{125 \times 6 + 140 \times 3 + 116 \times 1}{6 + 3 + 1} = 128 \cdot 6$$

In general, if there are $n$ commodities with base-year prices $p_0'$, $p_0''$, ... $p_0^n$ and current prices $p_1'$, $p_1''$, ... $p_1^n$, the simple or unweighted AM is (omitting the factor 100):

$$\frac{1}{n}\left\{\frac{p_1'}{p_0'} + \frac{p_1''}{p_0''} + \ldots + \frac{p_1^n}{p_0^n}\right\} = \frac{1}{n}\Sigma\left(\frac{p_1}{p_0}\right) \tag{17.1}$$

and the weighted AM with weights $w'$, $w''$, ... $w^n$ is

$$\frac{1}{\Sigma w}\left\{w'\frac{p_1'}{p_0'} + w''\frac{p_1''}{p_0''} + \ldots + w^n\frac{p_1^n}{p_0^n}\right\} = \frac{1}{\Sigma w}\Sigma\left(w\frac{p_1}{p_0}\right) \tag{17.2}$$

In these expressions the superscripts denote the several commodities and the subscripts the years concerned, e.g. $p_0^n$ denotes the price of the *n*th commodity in the base year, not a base-year price raised to the *n*th power.

A disadvantage of the AM with fixed weights is that it gives too much weight to items that have risen in price, i.e. there is an upward bias in it. When a commodity has suffered a great increase in price, it is generally because it has become scarce, and when it becomes cheap it is generally because it is plentiful. One solution is to vary the weighting, but this method also has its drawbacks, and it may be difficult to decide what weights to use.

## The geometric mean

The geometric mean of *n* price relatives is simply the *n*th root of their product (see page 101), and it is easily found with the aid of logarithms. The formula for the simple or unweighted GM is:

$$P = \sqrt[n]{\left(\frac{p_1'}{p_0'} \times \frac{p_1''}{p_0''} \times \ldots \times \frac{p_1^n}{p_0^n}\right)} \tag{17.3}$$

or, taking logarithms,

$$\log P = \frac{1}{n} \Sigma \log \frac{p_1}{p_0} \tag{17.4}$$

*Example* 17.2

Calculate a price index for 1976 (1975 = 100) for the data of Example 17.1, using the geometric mean.

| Commodity | Price relative | Log | Anti-log |
|-----------|----------------|-----|----------|
| A | 125 | 2·0969 | |
| B | 140 | 2·1461 | |
| C | 116 | 2·0645 | |
| Total | | 3)6·3075 | |
| | | 2·1025 | 126·6 |

The answer, 126·6, is slightly less than the AM, 127·0, as it must be, but the difference is not very material. Indeed, it might be argued that the AM and GM would only differ appreciably if the price relatives varied so widely that a price index had little meaning.

It is not really necessary to calculate each price relative with the

GM as it is with the AM, since formulae (17.3) and (17.4) may be expressed as:

$$P = \sqrt[n]{\left(\frac{p_1' p_1'' \cdots p_1^n}{p_0' p_0'' \cdots p_0^n}\right)} \tag{17.3a}$$

and

$$\log P = \frac{1}{n}(\Sigma \log p_1 - \Sigma \log p_0) \tag{17.4a}$$

In the GM, as in the AM, items can be weighted according to their importance, although the weights are no longer factors but indices in the usual algebraic sense of the word. The weighted GM of $n$ items with weights $w', w'', \ldots w^n$ is given by:

$$P = \sqrt[(\Sigma w)]{\left[\left(\frac{p_1'}{p_0'}\right)^{w'} \left(\frac{p_1''}{p_0''}\right)^{w''} \cdots \left(\frac{p_1^n}{p_0^n}\right)^{w^n}\right]} \tag{17.5}$$

or

$$\log P = \frac{1}{\Sigma w} \Sigma \left(w \log \frac{p_1}{p_0}\right) \tag{17.6}$$

The principle is really the same in both cases: an item with a weight of, say, 15, is treated as if it were 15 separate items.

Obviously the GM is more troublesome to compute than the AM and not so easy to interpret, but it has certain theoretical advantages over the AM. Before considering these further, however, it will be convenient to deal with the method of aggregates.

## The aggregative method

This is the best method to use if the data are available. It consists, as already mentioned, in taking a 'bundle' of goods and comparing their value at current prices with their value at base prices. The bundle is generally one of two things:

(i) the quantities $q_0', q_0'', \ldots q_0^n$ actually bought or sold in the base year; or

(ii) the quantities $q_1', q_1'', \ldots q_1^n$ actually bought or sold in the current year (or month etc.).

In the first case the index measures the change in cost of a fixed bundle of goods. The formula, named after Laspeyres, is:

$$P = \frac{q_0'p_1' + q_0''p_1'' + \ldots + q_0''p_1''}{q_0'p_0' + q_0''p_0'' + \ldots + q_0''p_0''} = \frac{\Sigma q_0 p_1}{\Sigma q_0 p_0} \qquad (17.7)$$

$$= \frac{\text{base-year quantities at current prices}}{\text{base-year quantities at base-year prices}}$$

In the second case the bundle varies from year to year or from month to month, and it is possible for the index to change even when all prices are unchanged. The formula, named after Paasche, is as follows:

$$P = \frac{q_1'p_1' + q_1''p_1'' + \ldots + q_1''p_1''}{q_1'p_0' + q_1''p_0'' + \ldots + q_1''p_0''} = \frac{\Sigma q_1 p_1}{\Sigma q_1 p_0} \qquad (17.8)$$

$$= \frac{\text{current quantities at current prices}}{\text{current quantities at base-year prices}}$$

It does not matter whether all the prices are in the same units as long as the $pq$ products are. Nor, of course, does it matter whether we write $\Sigma q_0 p_1$ or $\Sigma p_1 q_0$.

*Example* 17.3

Calculate Laspeyres and Paasche price indices for 1976 (1973 = 100) from the following sales data:

| Commodity | Sales in 1973 | | Sales in 1976 | |
|---|---|---|---|---|
| | Tons | £ | Tons | £ |
| A | 200 | 6,000 | 500 | 20,000 |
| B | 420 | 10,500 | 400 | 14,000 |
| C | 2,000 | 20,000 | 1,500 | 30,000 |
| D | 5 | 3,500 | 45 | 36,000 |
| Total | | 40,000 | | 100,000 |

To do this we must calculate the average prices for each year and compile a second table.

| Commodity | $p_{73}$ | $q_{76}$ | $p_{73}q_{76}$ | $p_{76}$ | $q_{73}$ | $p_{76}q_{73}$ |
|---|---|---|---|---|---|---|
| A | 30 | 500 | 15,000 | 40 | 200 | 8,000 |
| B | 25 | 400 | 10,000 | 35 | 420 | 14,700 |
| C | 10 | 1,500 | 15,000 | 20 | 2,000 | 40,000 |
| D | 700 | 45 | 31,500 | 800 | 5 | 4,000 |
| Total | | | 71,500 | | | 66,700 |

The totals $\Sigma p_{73}q_{73}$ and $\Sigma p_{76}q_{76}$ are, of course, already calculated.

The Laspeyres price index is:

$$P_L = \frac{66,700}{40,000} \times 100 = 167$$

and the Paasche price index is:

$$P_P = \frac{100,000}{71,500} \times 100 = 140$$

Calculations for a series of years are simpler with the Laspeyres formula, since the denominator, being the total volume for the base year, is constant and need only be calculated once.

These aggregative formulae are equivalent to the weighted AM of price relatives, with suitably chosen weights. Thus the Laspeyres formula can be written:

$$P = \frac{p_0'q_0'\left(\dfrac{p_1'}{p_0'}\right) + p_0''q_0''\left(\dfrac{p_1''}{p_0''}\right) + \ldots + p_0{}^n q_0{}^n \left(\dfrac{p_1{}^n}{p_0{}^n}\right)}{p_0'q_0' + p_0''q_0'' + \ldots + p_0{}^n q_0{}^n} \qquad (17.7a)$$

which is the weighted AM of $p_1'/p_0'$ etc. with weights $p_0'q_0'$, $p_0''q_0''$ etc. These products are the actual values of sales of the various items in the base year. Similarly, the Paasche formula is the AM of price relatives with weights $p_0'q_1'$ etc., the values of current sales at base-year prices.

## Index numbers of quantity

Indices of quantity or volume present their own problems. There is no difficulty with a fairly homogeneous product like coal or steel, but what is meant, for instance, by the volume of production of the chemical industry? Obviously, not simply the physical volume of the many items involved. To add together heavy chemicals, dyes, plastics and drugs and express the total in tonnes or litres would be ludicrous. Somehow these things must be weighted according to their importance. A tonne of penicillin must carry far more weight in the statistical sense than a tonne of soda ash.

The method usually employed is to calculate quantity relatives for each of the commodities included in the index, and take their weighted arithmetic mean, the weights being roughly proportional to the value of production, sales or whatever may be more appropriate, in the base year. Suppose, for example, that we require an index of sales volume. Then, with the same notation as before, and denoting the volume index by $Q$,

$$Q = \frac{1}{\Sigma w} \left\{ w' \frac{q_1'}{q_0'} + w'' \frac{q_1''}{q_0''} + \ldots + w^n \frac{q_1^n}{q_0^n} \right\} = \frac{1}{\Sigma w} \Sigma \left( \frac{q_1}{q_0} \right)$$

(17.9)

If the weights are exactly equal to the base-year values, so that $w' = q_0' p_0'$ etc., formula (17.9) becomes

$$Q = \frac{q_1' p_0' + q_1'' p_0'' + \ldots + q_1^n p_0^n}{q_0' p_0' + q_0'' p_0'' + \ldots + q_0^n p_0^n} = \frac{\Sigma q_1 p_0}{\Sigma q_0 p_0}$$

(17.10)

We have thus obtained our index as the ratio of two total values, one of current quantities, the other of base-year quantities, both valued at base-year prices.

It may be objected that the purpose of the index is to measure physical volume, not value. The answer to that is (i) that a multitude of dissimilar things cannot be added together in any sensible unit except a unit of money (after all, money was devised for the very purpose of providing a common unit), (ii) that the main objection to an index based on value is that the value of money itself varies, and this objection does not apply to an index of value at fixed prices.

An alternative but generally less meaningful index of volume would be the ratio of total values at current prices, i.e.

$$Q = \frac{\Sigma q_1 p_1}{\Sigma q_0 p_1}$$

(17.11)

Formulae (17.10) and (17.11) might be called the Laspeyres and Paasche quantity indices respectively. If we denote them by $Q_L$ and $Q_P$, and similarly for the price indices, we see that

$$P_L \times Q_P = \frac{\Sigma p_1 q_0}{\Sigma p_0 q_0} \times \frac{\Sigma q_1 p_1}{\Sigma q_0 p_1} = \frac{\Sigma p_1 q_1}{\Sigma p_0 q_0}$$

and that

$$P_P \times Q_L = \frac{\Sigma p_1 q_1}{\Sigma p_0 q_1} \times \frac{\Sigma q_1 p_0}{\Sigma q_0 p_0} = \frac{\Sigma p_1 q_1}{\Sigma p_0 q_0}$$

so that both products are equal to the index of total value. Our quantity index could, in fact, have been obtained by dividing the index of total value by the price index, or vice versa. One, however, would be of the Laspeyres type and the other of the Paasche type, and it is unlikely that both would be ideal for their purpose. In other words, we cannot have it both ways.

*Example* 17.4

Calculate Laspeyres and Paasche indices of sales volume from the data of Example 17.3.

$$Q_L = \frac{\Sigma q_{76} p_{73}}{\Sigma q_{73} p_{73}} = \frac{71,500}{40,000} \times 100 = 179$$

and

$$Q_P = \frac{\Sigma q_{76} p_{76}}{\Sigma q_{73} p_{76}} = \frac{100,000}{66,700} \times 100 = 150$$

The reader should verify that he gets the same results by taking the index of total value, which is 250, and dividing by the appropriate price index as found in Example 17.3.

### Hybrid formulae

The Laspeyres formula, like its equivalent, the weighted AM of price relatives, tends to overstate a rise in prices because it takes no account of the fall in output or demand that often accompanies a marked increase in price, or of the expansion of production that causes a fall in price. An item may be priced right out of the market, but it retains the weight appropriate to a normal price. For similar reasons, the Paasche index goes to the other extreme and tends to understate the rise in prices.

The two formulae may be regarded as giving upper and lower limits, the true index (if there is such a thing) being somewhere between them. Various compromises have been suggested. One is to average or add together the base-year and current-year quantities, giving the formula:

$$P = \frac{\Sigma p_1 (q_0 + q_1)}{\Sigma p_0 (q_0 + q_1)} \tag{17.12}$$

Another is to take the AM or the GM of the Laspeyres and Paasche indices. In practice, the difference between the AM and the GM is negligible, but the GM is preferred because, as will be shown later, it satisfies certain mathematical tests. The formula is:

$$P = \sqrt{\left(\frac{\Sigma q_0 p_1}{\Sigma q_0 p_0} \times \frac{\Sigma q_1 p_1}{\Sigma q_1 p_0}\right)} \tag{17.13}$$

This is called Fisher's 'Ideal' formula, after its chief exponent, the late Professor Irving Fisher (not to be confused with the British statistician, the late Professor Sir Ronald Fisher).

The principal formulae for calculating index numbers have now been described, viz. the AM and the GM of relatives, simple or weighted, the Laspeyres and Paasche aggregative formulae, and combinations or variations of them. Before considering their relative merits in detail, it will be necessary to describe the Chain-base Method and the theoretical tests mentioned above.

## The chain-base method

Up to now it has been assumed that each year is compared directly and independently with the base year. Sometimes, however, it is preferable to compare each year with the previous year and build up the resulting 'links' into a chain by successive multiplication. This is called the Chain-base Method. Suppose, for example, the index for 1975 based on 1974 is 112·5 and the index for 1976 based on 1975 is 105·6; then the index for 1976 based on 1974 is:

$$112 \cdot 5 \times \frac{105 \cdot 6}{100} = 118 \cdot 8$$

This method provides a more direct comparison between successive years than the Fixed-base Method and makes it easy to change the base year when desired. Its chief advantage is that it takes account of the changing pattern of sales etc. and, in particular, it enables new products to be introduced into the index as soon as there are two years' prices to compare; it also enables old products to be retained as long as possible. With the Fixed-base Method, only products common to the current year and the base year could be used. Reclassification of products also creates difficulties, but if both classifications are available for the year of transition, it can be linked up with both the previous year and the following year, giving a continuous index.

A disadvantage of the Chain-base Method is that any defect or abnormality in the index for one year is perpetuated in all subsequent years; the chain is only as strong as its weakest link.

### Tests for consistency

As has been noticed several times in this chapter, certain index-number formulae are not entirely self-consistent, and several theoretical tests have been suggested for judging the suitability of these formulae. Two such tests can be applied to index numbers relating two years with one another; these are:

(i) the time-reversal test;
(ii) the factor-reversal test.

A third test, called the circular test, concerns the relations between three or more years. These three tests will be discussed in turn.

1. The *time-reversal test* implies that a price index, like an individual price, should show the same relative movement from one year to another whichever year is taken as base, i.e. that if $P_{12}$ denotes the index for Year 2 based on Year 1, and $P_{21}$ that for Year 1 based on Year 2,

$$P_{12}: 100 = 100: P_{21}$$

or, disregarding the factor 100,

$$P_{12}P_{21} = 1 \tag{17.14}$$

To take an extreme case, which would not occur in practice, suppose there are only two commodities $A$ and $B$, equally weighted, and that the price of $A$ remains constant, while the price of $B$ in Year 2 is four times its price in Year 1. Then, with Year 1 as base, the relatives are 100 and 400, and with Year 2 as base, they are 100 and 25.

Now with the AM,

$$P_{12} = \tfrac{1}{2}(100 + 400) = 250$$
$$P_{21} = \tfrac{1}{2}(100 + 25) = 62 \cdot 5$$

whereas with the GM,

$$P_{12} = \sqrt{(100 \times 400)} = 200$$
$$P_{21} = \sqrt{(100 \times 25)} = 50$$

Clearly, the GM is time-reversible and the AM is not.

2. The idea behind the *factor-reversal test* is that a formula involving both prices and quantities, if right for a price index, should be right for a quantity index when $p$'s and $q$'s are interchanged, and that the product of the two indices should be equal to $V$, the index of value.

This does not apply to any formula which contains $p$'s only. If the $p$'s and $q$'s are interchanged in the Laspeyres formula,

$$P = \frac{\Sigma p_1 q_0}{\Sigma p_0 q_0},$$

$$Q = \frac{\Sigma q_1 p_0}{\Sigma q_0 p_0},$$

and

$$PQ = \frac{(\Sigma p_1 q_0)(\Sigma q_1 p_0)}{(\Sigma p_0 q_0)^2},$$

which is not equal to $V$, so that the test is not satisfied.

Similarly with the Paasche formula. In fact, the only formula given in this chapter that satisfies the factor-reversal test is the 'Ideal' formula.

From formula (17.13),

$$P = \sqrt{\left(\frac{\Sigma p_1 q_0}{\Sigma p_0 q_0} \times \frac{\Sigma p_1 q_1}{\Sigma p_0 q_1}\right)}$$

and interchanging $p$'s and $q$'s,

$$Q = \sqrt{\left(\frac{\Sigma q_1 p_0}{\Sigma q_0 p_0} \times \frac{\Sigma q_1 p_1}{\Sigma q_0 p_1}\right)}$$

$$PQ = \sqrt{\left(\frac{\Sigma p_1 q_0}{\Sigma p_0 q_0} \times \frac{\Sigma p_1 q_1}{\Sigma p_0 q_1} \times \frac{\Sigma p_0 q_1}{\Sigma p_0 q_0} \times \frac{\Sigma p_1 q_1}{\Sigma p_1 q_0}\right)}$$

$$= \frac{\Sigma p_1 q_1}{\Sigma p_0 q_0} = V,$$

and the test is satisfied.

3. The *circular test* is an extension of the time-reversal test to more than two years. If $P_{12}$, $P_{23}$ etc. are indices (omitting the factor 100) for Year 2 based on Year 1, for Year 3 based on Year 2, and so on, the circular test requires that:

$$P_{12}P_{23} \ldots P_{n-1,n} P_{n1} = 1 \tag{17.15}$$

When $n = 2$ this equation reduces to formula (17.14). It follows that $P_{1n}P_{n1} = 1$, and consequently that

$$P_{1n} = P_{12}P_{23} \ldots P_{n-1,n}$$

The circular test is thus more stringent than the time-reversal test, as it includes the latter as a special case, and it also implies that, under any formula satisfying the test, the chain-base and fixed-base methods give the same result.

One formula that satisfies the circular test is the aggregative formula with fixed weights, i.e.

$$P = \frac{\Sigma wp_1}{\Sigma wp_0} \tag{17.16}$$

For simplicity, this will be shown for three years only:

$$P_{12}P_{23}P_{31} = \frac{\Sigma wp_1}{\Sigma wp_0} \times \frac{\Sigma wp_2}{\Sigma wp_1} \times \frac{\Sigma wp_0}{\Sigma wp_2} = 1$$

Another formula that satisfies the circular test is the geometric mean, unweighted or with fixed weights.

Fisher's 'Ideal' formula, for which somewhat extravagant claims have been made by some of its advocates, fails to satisfy the circular test.

## What makes a good index number?

It might be thought, from the amount of space devoted to tests for consistency in this chapter and in other literature on index numbers, that they gave a perfect criterion for a good formula. In the author's opinion, however, there are other things at least equally important. First, an index number should be simple in conception and easily interpreted. The Laspeyres formula scores heavily here: the man in the street can understand an index that tries to measure the changing cost of the things he bought in a particular year. The Paasche formula is also easily understood, and so is the AM of price relatives, but the GM is more difficult to interpret. So is the 'Ideal' formula, except as a compromise between two formulae that are biased in opposite directions.

Secondly, an index should be reliable—not necessarily precise, but reasonably accurate for its purpose. An unweighted AM or an AM with fixed weights is liable to give misleading results—generally too high—as time goes on and the weighting becomes obsolete. The GM is an attempt to correct the upward bias of the AM and, in practice, the GM with base-year weights is fairly accurate for small movements, but an unweighted GM or AM is a confession of ignorance, for it suggests that the compiler has no idea what the correct weighting is.

Thirdly, an index number should be reasonably consistent. This is really a corollary of the second point, for if it is grossly inconsistent it cannot be accurate. For example, the effect of changing the base year, and revising the weighting at the same time, should be to make slight, not drastic, changes in relative movements of the index.

The author does not take the view that an index that satisfies the various mathematical tests is therefore the right index. Indeed, there is no such thing as a single 'right index', appropriate in all circumstances. The Sales Manager may say to the statistician, 'Our sales have doubled in value in the last ten years. How much of that is increase in prices and how much is increase in volume?' As Examples 17.3 and 17.4 have shown, there is no simple and unique answer.

High claims have been made by many people for Fisher's 'Ideal' formula, but it is rarely used because:

(*a*) it involves twice as much work as the Laspeyres or Paasche formula separately;

(*b*) it is not quite so easy to interpret;

(*c*) current quantities (or in the case of an index of quantity, current prices) are often unknown; and

(*d*) even the 'Ideal' formula is not consistent for a series of years, but only between the old base year and the new. In fact, if conditions are constantly changing, necessitating frequent revision of the weighting, it is impossible to satisfy all tests of consistency at the same time.

This raises an important distinction between long-term and short-term movements. In a monthly series covering a few years only, especially if there are marked seasonal movements, fixed weights are generally best for, with current weighting, it would be possible for variations in the pattern of sales, imports etc. to cause fluctuations in the index even when prices remained constant. Fixed weights ensure that changes in the index really do reflect changes in prices. Over a long period, however, changes in pattern must be allowed for. This can be done either by using current weighting or by running a fixed weight index for a few years, then revising the weighting and linking up the two series.

## Exercises

17.1 (*a*) What are the main factors to be taken into consideration in constructing an index number? Relate your observations specifically to an index of retail prices.

(*b*) For the share portfolio given below, construct index numbers for 1971 and 1972 using 1970 as the base. You are to assume that the valuations are taken from the same source on the same day in subsequent years and that no transfer of shares has taken place:

| Class of share | Number of shares | Price per share | | |
|---|---|---|---|---|
| | | 1970 (£) | 1971 (£) | 1972 (£) |
| Industrials | 350 | 1·60 | 1·80 | 2·00 |
| Property | 200 | 4·60 | 5·00 | 6·50 |
| Consumer durables | 120 | 0·50 | 0·70 | 0·60 |
| Food and clothing | 150 | 0·40 | 0·40 | 0·50 |
| Miscellaneous | 160 | 1·00 | 0·90 | 0·50 |

(*ICMA*)

17.2

Clothing and Footwear:

| Item | Weights | Price Relatives | |
|---|---|---|---|
| | | Dec. 1972 Jan. 1972 = 100 | Feb. 1974 Jan. 1974 = 100 |
| Men's outer clothing | 16 | 159 | 101 |
| Men's under clothing | 6 | 156 | 104 |
| Women's outer clothing | (*a*) | 146 | (*c*) |
| Women's under clothing | 4 | 143 | 104 |
| Children's clothing | 11 | 142 | 104 |
| Hose, hats, etc. | 12 | 132 | 101 |
| Footwear | 18 | 153 | 101 |
| Total | 91 | (*b*) | 102·6 |

For the table above find (*a*), (*b*) and (*c*). Explain the terms 'Weights' and 'Price Relatives'.　　　　　　　　　　　　　　　　　　　(*IoS*)

17.3 From the following data, calculate an index number for measuring the change in the general level of prices of the group of goods listed:

| Commodity | Unit of purchase | Prices | | Quantities | |
|---|---|---|---|---|---|
| | | Period 0 $(p_0)$ pence | Period 1 $(p_1)$ pence | Period 0 $(q_0)$ | Period 1 $(q_1)$ |
| Bread | 2 lb | 18 | 21 | 100 | 90 |
| Tea | 1 lb | 80 | 90 | 10 | 8 |
| Potatoes | 7 lb | 49 | 35 | 20 | 30 |
| Butter | 1 lb | 56 | 56 | 30 | 30 |
| Milk | Quart | 21 | 18 | 7 | 10 |
| Beef | 1 lb | 42 | 63 | 80 | 60 |
| Mutton | 1 lb | 30 | 40 | 70 | 60 |

Give reasons for the method of index number calculation chosen.

(*ACA*)

17.4 Discuss the relative advantages and disadvantages of Laspeyres and Paasche indices. For the following data calculate Paasche and Laspeyres price index numbers:

| | 1969 | | 1974 | |
|---|---|---|---|---|
| | Price per unit (New pence) | Quantity No. of units | Price per unit (New pence) | Quantity No. of units |
| Beef | 35 | 3 | 80 | 2 |
| Potatoes | 2 | 2 | 4·5 | 3 |
| Peas | 5 | $1\frac{1}{2}$ | 9 | 1 |

(*IoS*)

17.5 Five feed components are to be used in the construction of an animal feed stuff index number. From the figures tabulated below, calculate:

(i) A Laspeyres price index.
(ii) A Paasche quantity index, taking 1964 = 100 in each case.

| Component | 1964 | | 1970 | |
|---|---|---|---|---|
| | Price/Ton (£) | Consumption (tons) | Price/Ton (£) | Consumption (tons) |
| A | 40 | 3000 | 41 | 2750 |
| B | 39 | 2750 | 53 | 1500 |
| C | 38 | 2050 | 35 | 2350 |
| D | 37 | 500 | 30 | 750 |
| E | 36 | 1475 | 24 | 2850 |

(*IoS*)

17.6   The following figures relate to all the exports of a certain country over a period of 10 years:

| Year | Value of exports in £ million based on | | |
|---|---|---|---|
| | (1) *Actual prices and quantities exported each year* | (2) *Actual quantities exported but prices in year 1* | (3) *Actual prices but quantities exported in year 1* |
| 1 | 100 | 100 | 100 |
| 2 | 107 | 98 | 110 |
| 3 | 112 | 97 | 119 |
| 4 | 115 | 94 | 132 |
| 5 | 116 | 91 | 140 |
| 6 | 115 | 90 | 150 |
| 7 | 112 | 89 | 162 |
| 8 | 107 | 86 | 171 |
| 9 | 100 | 83 | 179 |
| 10 | 91 | 82 | 191 |

(*a*)   Column 2 provides an index for quantities exported each year, using year 1 as base. Calculate an alternative index for quantities, using year 1 as base.

(*b*)   Compare your own index with that given in the table above, explaining why any differences may have arisen. State, with reasons, which type of index number you prefer here.          (*IoS*)

17.7   Production and prices of a country's three main vegetables in the years 1962, 1963 and 1964 are given below:

| Vegetable | 1962 | | 1963 | | 1964 | |
|---|---|---|---|---|---|---|
| | *Thousand tons* | *Price/ton (£)* | *Thousand tons* | *Price/ton (£)* | *Thousand tons* | *Price/ton (£)* |
| Potatoes | 5,000 | 3 | 6,000 | 2 | 7,000 | 3 |
| Cabbages | 4,000 | 6 | 4,000 | 6 | 5,000 | 5 |
| Turnips | 8,000 | 4 | 9,000 | 6 | 12,000 | 7 |

Construct two different indices of physical volume. Quote the formulae used in their construction and state briefly the differences in meaning and use between the two indices.          (*IoS*)

17.8   Explain the difference between Laspeyres and Paasche price index numbers.

In a Laspeyres index designed to measure the changes in price of dairy products, the price of a certain variety of cheese in 1976 is wrongly recorded as £0.70 per lb, and the value of the index calculated as 125. The correct price is £0.60 per lb, as compared with £0.50 per lb in 1970, the base year. If the ratio of the annual expenditure on this cheese to the total expenditure on dairy products dropped from 6·0 per cent in 1970 to 4·2 per cent in 1976, what is the correct value of the index for 1976? If the value of 125 in 1976 had been a Paasche index, what would be the correct value? (*IoS*)

# 18 Introduction to published statistics

The remaining chapters of this book deal almost entirely with the principal sources, methods of compilation, and interpretation of published statistics, i.e. data published by official bodies or private bodies with a recognized status. Not only will this give the student a working knowledge of the statistics themselves, but it will illustrate many of the principles expounded in earlier chapters. No attempt is made to cover the whole field, and many topics such as finance, education and crime are omitted. As a general rule, only British statistics are considered, and statistics and statistical publications for Wales, Scotland and Northern Ireland will not be mentioned separately.

The reader must remember that this book is going to the printer in 1977, and that, although every effort has been made to give up-to-date information, by the time he reads these chapters there will almost certainly be changes in the scope and content of some of the statistics described. Some of the historical series will have been revised also.

Before considering the publications themselves, it is desirable to give a brief account of the background and development of official statistics in the United Kingdom.

## The development of British official statistics

Like Topsy in *Uncle Tom's Cabin*, official statistics 'just growed', slowly and painfully, as a by-product of administrative requirements. Before the nineteenth century they were virtually non-existent, and estimates of population, production etc. were extremely uncertain. At one point in our history it was commonly thought that the sex ratio was about 2:1, i.e. two females to one male. However, a step towards reliable demographic statistics was taken in 1801 with the first Census of Population (see Chapter 19). Statistics of the country's economy were much later developing: there was no Census of Production until 1907.

The inter-war years saw considerable progress both in statistical theory and its applications, and in the development of economic statistics, together with a growing awareness of the inadequacy of the available data. The Second World War highlighted the need for reliable figures and for better organization, and in 1941 the Central Statistical Office (CSO for short) was set up to co-ordinate the statistical work of the various Government Departments. For instance, at that time the Board of Trade, the Ministry of Labour and other Government Departments all used different classifications of industry, and it was difficult to relate, say, the figures of employment and unemployment to the statistics given in the Census of Population or the Census of Production. In 1948, therefore, the CSO issued a Standard Industrial Classification in order to secure uniformity and comparability in the statistics of the various Departments. It is now used for the Censuses of Population, Production and Distribution, and other official statistics.

After the war, published statistics expanded rapidly, but never quite fast enough. In 1956, Harold Macmillan, then Chancellor of the Exchequer, made his famous speech complaining of the deficiencies in the available statistics and the delay in publishing them ('We are always, as it were, looking up a train in last year's Bradshaw'). After that, there was a considerable expansion in statistics of national income, balance of payments, production, distribution, stocks, capital expenditure etc., followed a few years later by more adequate statistics of hitherto neglected topics such as finance and housing. Many of these figures were by now published quarterly as well as annually, and regional analyses became more and more common, following the development of a system of Standard Regions.

The next big step forward came in 1969, when the Business Statistics Office (BSO) was formed to centralize and integrate the various censuses and enquiries relating to production, sales and other aspects of business. These are further discussed in Chapters 22 and 23. The aim of the BSO is to serve business as well as government. One gigantic task was to compile a Central Register of Businesses to replace the separate registers held by different Government Departments and provide a sampling frame for the growing number of sample enquiries.

Almost concurrently with this, the General Register Office and the Government Social Survey merged in a new Office of Population Censuses and Surveys (OPCS), which is responsible for the Census of Population and various annual or continuous surveys such as the

General Household Survey (see Chapter 19). The BSO and the OPCS are now the principal collecting agencies. To sum up, the Government Statistical Service now comprises the statistics divisions of all major departments, together with the CSO, the BSO and the OPCS.

## British statistical publications

Most Government Departments publish annual reports, and some publish their own journals, the best known being the weekly *Trade and Industry* and the monthly *Department of Employment Gazette*. The most important statistical publications, however, are compiled by the Central Statistical Office, the chief of them being:

1. *The Monthly Digest of Statistics* which, as its name indicates, is an official summary of statistics published each month, containing about 180 tables on a wide range of subjects. An annual supplement, entitled *Definitions and Explanatory Notes*, provides a useful account of the tables shown and units employed in the *Monthly Digest*.

2. *Economic Trends*, a monthly booklet of small charts and tables of key statistics designed to show the principal trends in the economic life of the United Kingdom. It is largely a condensed pictorial version of the *Monthly Digest*, but there are also special articles in each issue dealing with specific aspects of economic statistics. The *Economic Trends Annual Supplement* provides notes and definitions and contains long runs of annual and quarterly data.

3. The *Annual Abstract of Statistics*, covering much the same ground as the *Monthly Digest* and a good deal more, containing about 400 tables. With a few exceptions it gives only annual figures and usually covers a period of at least ten years.

4. *National Income and Expenditure of the United Kingdom*, an annual Blue Book which will be discussed at length in Chapter 25.

5. *Statistical News,* a quarterly booklet designed to keep industry and the public informed about current developments.

6. *Social Trends*, an annual selection of tables and charts designed to illustrate trends and patterns in society, e.g. population, housing, health etc.

7. *Financial Statistics*, a monthly booklet giving the main financial and monetary statistics of the United Kingdom.

8. *Regional Statistics*, published annually, containing about 80 tables giving analyses by Standard Region.

The CSO also publishes a series of *Studies in Official Statistics* on subjects such as the Index of Industrial Production and the Index of Retail Prices. It has recently issued an extremely comprehensive *Guide to Official Statistics*, a large reference book covering about 2500 sources, which has greatly helped the revision of this book.

The BSO publishes a very large number of *Business Monitors* to meet the needs of industry and commerce. Some are annual, others quarterly, and a few are monthly. There are three main series:

(*a*) Production Monitors (see Chapter 22).
(*b*) Service and Distributive Monitors (see Chapter 23).
(*c*) Miscellaneous Monitors on such varied topics as cinemas, company finance, overseas trade and tourism.

A considerable number of statistical yearbooks and bulletins are published by trade associations and other bodies, e.g.:

*World Metal Statistics*
*Iron and Steel Industry Monthly Statistics*
*The Motor Industry of Great Britain* (annual)
*Rubber Statistical Bulletin* (monthly)
*Lloyd's Shipping Register* (quarterly)

Statistical series and articles are also published by a number of unofficial or semi-official bodies, notably in the *National Institute Economic Review*, published quarterly by the National Institute of Economic and Social Research.

Finally, there is a mass of unpublished detail available to the public on request, often for a small fee. For example, the CSO holds a data bank of macro-economic time series, from which one may obtain information in the form of computer tape or print-outs.


## Suggestions for further reading

*N.B.* In this and subsequent chapters, references and suggestions for further reading are published by HMSO unless otherwise specified.

Berman, L. S. (1971). Recent Improvements in Official Economic Statistics. *J. Roy. Stat. Soc.*, Series A, **134,** Part 4.

Edwards, B. *Sources of Economic and Business Statistics*. London: Heinemann.

Fessey, M. C. (1969). Some Developments in Economic Statistics since 1934. *The Statistician*, **19,** No. 2.

Harvey, J. M. *Sources of Statistics*. Clive Bingley.

Lewes, F. M. M. (1971) *Statistics of the British Economy.* London: Allen & Unwin.

*Economic Trends* (M)

*Guide to Official Statistics*

*Studies in Official Statistics,* Nos. 5, 9, 10, 12, 15, 19 and 24. *New Contributions to Economic Statistics.* (Articles reprinted from *Economic Trends.*)

*Statistical News* (Q)

*Social Trends* (A)

# 19 Demographic statistics

Demography is the study of people from a statistical point of view and deals with such matters as population, births and deaths, marriages and fertility. It will also be taken here to include sickness or, as it is sometimes called, morbidity, but employment and income statistics will be left to later chapters.

Accurate demographic statistics require two things: a periodic census, or stocktaking of the whole population at a given time; and a system of compulsory registration of births, deaths and marriages. In England and Wales, the Director and Registrar General, as head of the Office of Population Censuses and Surveys (incorporating the General Register Office), is responsible for these functions. Scotland and Northern Ireland have their own Registrar-Generals and General Register Offices and publish separate reports, but the systems are very similar to that of England and Wales and there is close co-ordination between them.

The General Register Office at Somerset House and what is substantially the present system of registration were set up by an Act of 1836. Registration was compulsory from the outset, although penalties for failure to register were not imposed until 1874. In 1874 also, the issue of medical certificates of cause of death was made compulsory, which greatly facilitated the analysis of deaths by cause.

## The census of population

Necessary as registration is, the basis of all population estimates is the census. The first census in Great Britain was taken in 1801, and from then until 1931 inclusive, a census was taken every tenth year, generally with slight differences between the census for England and Wales and those for Scotland and Northern Ireland. Owing to the war, there was no census for 1941, although the National Register taken in

September 1939 helped to fill the gap. The decennial census was resumed in 1951, and is being continued, but a 10 per cent sample census was also held in 1966.

The population recorded is *de facto*, not *de jure*. In plain English, this means that a man is recorded for the district in which he happens to be at the time, not necessarily that in which he normally lives. It is therefore desirable to avoid holiday seasons and other times when many people are away from home. The time usually chosen is midnight on a Sunday in April. The population of the country as a whole is also *de facto*; thus it excludes people abroad on business or on holiday and British soldiers serving overseas, but includes American service personnel stationed in England and other foreign visitors.

One problem still to be overcome is the inevitable delay in converting millions of schedules into statistical tables and publishing the results. The use of high-speed tabulating machines in the 1911 and later Censuses did not greatly reduce this delay, as it stimulated the demand for more detailed tabulations, and not only do the schedules have to be examined and coded as before, but fifty million cards have to be punched by hand before they can be sorted and tabulated. Much of the editing, tabulation and analysis is now done by computer, but the preliminary work still has to be done by hand. However, it is now possible to obtain from the census office (for a small fee) special tabulations or computer tape derived from census data, long before the reports are published.

Every effort is made to ensure accuracy in the census, but some errors are unavoidable. They are chiefly errors of age, marital status and occupation. Uncertainty will cause a man to enter his age as 70 rather than 69; there is a distinct bias towards ages ending in 0, and a smaller bias towards ages ending in 8. Vanity may induce a woman of 45 to give her age as 40, a divorced woman to describe herself as widow, and a clerk to give his occupation as 'official' or 'executive'. Although the standard of enumeration for the population in general is high, there is a tendency to understate the population of certain groups who present special difficulties. This is especially true of the mobile element, for example, vagrants and people living in caravans.

Improvements in technique and organization, however, are steadily reducing the number of errors made. The 1961 Census was followed by the first 'post-enumeration survey' by trained field staff to check errors of enumeration and response. One result of this is that census forms now ask for date of birth instead of age, thereby greatly reducing errors of age.

The 1966 Census was unusual, not only because it was held five years after the previous census, but also because it was taken on a 10 per cent sample of households. One of its main objects was to obtain information about housing, so there were questions about number of rooms, whether the household had a hot-water supply, bath, WC etc., number of cars owned and where they were kept. There was also a new question about mode of transport to work.

At the time of writing, the results of the 1971 Census are still being published. About 100,000 enumerators were required to deliver and later to collect the forms. There were special forms for hotels and boarding houses, ships etc., but the principal form was the 'H Form for Private Households'. There were three parts to the H Form, which may be summarized as follows:

*Part A* contained questions about the household's accommodation—type of tenure, number of rooms, number of cars and vans, and use of baths, toilets, cookers etc.

*Part B* asked for details of all persons present on Census night (25 April) or who joined the household on Monday, 26 April and had not been enumerated elsewhere, viz. sex, date of birth, marital status, country of birth (and country of birth of each person's father and mother), whether employed or retired etc., usual address one year ago and five years ago, educational qualifications, occupation, and means of transport to work. There were also questions about date of marriage and dates of birth of children, for women under 60 who were or had been married.

*Part C* asked for brief particulars of any person who usually lived in the household but who was not present and for whom, therefore, no entry had been made in Part B. The information so collected was used only in the Household Composition Tables.

## Birth rates and fertility

The *birth rate* for a given year is obtained by dividing the number of live births by the average population during the year and multiplying by 1000. In practice, the estimated population at the end of June is taken as the average for the year. For a shorter period, such as a quarter, the number of births must first be multiplied by four, or the appropriate factor, to give the equivalent annual rate.

The birth rate, although useful over periods of a few years, is not a good measure of the way the population is replacing itself, because so much depends on its age-distribution. Given normal fertility, a young population will produce a higher birth rate than a population with a high proportion of old people. A more useful measure, if the data are available, would be the *fertility rate*, obtained by relating total live births to the number exposed to risk, i.e. the number of women of reproductive age. This is bound to be somewhat arbitrary, but is generally taken as covering ages 15 to 44 inclusive. It is usual to distinguish legitimate and illegitimate fertility rates, but the former includes many births conceived before marriage.

*Reproduction rates* are different again, being built up from the age-specific fertility rates, i.e. the rates at each age of the potential mothers. Briefly, the *Gross Reproduction Rate* is the average number of daughters being born to women of reproductive age at current fertility rates. The *Net Reproduction Rate* takes mortality into account, being the average number of daughters being born who may be expected to survive to their mothers' ages and replace them as potential mothers. If the Net Reproduction Rate is unity, the population is replacing itself—or so it was once claimed.

It is now realized that Net Reproduction Rates may be misleading if, in the year in question, marriage and fertility rates were abnormal. In the years immediately following the Second World War there was a sharp rise in the number of births, owing to the large number of marriages that took place and the number of reunions between husband and wife who had not previously had children, or at least had had none for several years. For a reliable measure of replacement it is necessary to study each *cohort* of married women, i.e. those married in a particular year, and to analyse the fertility of each cohort by age at marriage and duration of marriage. From this *cohort analysis*, as it is called, it is possible to construct *General Replacement Rates* which take all these factors into account. Even then they are liable to fluctuate owing to temporary influences. The fact is that, as in economic affairs, no single index will suffice; there is no real substitute for a thorough examination of all the facts.

## Death rates

The *crude death rate* or *mortality rate* for a given year is the number of deaths per thousand, the equivalent annual rates being given for periods other than a year. Death rates can also be calculated for particular occupations, ages, or any other group of people exposed to risk. The death rate for a particular age or age-group is called an *age-specific death rate*. The *infant mortality rate* is the number of deaths of infants under one year old per 1000 live births. At the beginning of this century it was about 150 per 1000, but it is now about 20 per 1000.

Table 28 shows extracts from English Life Table No 12, 1960–1962 (Males). A life table is constructed from the age-specific mortality rates experienced by a given population during a given period (in this case by males in England and Wales during the years 1960–2), and shows what would happen to, say, 100,000 babies born if, in every future year of age, their mortality rates were those of the population observed. In Table 28, $l_x$ denotes the number who, on this assumption, would reach age $x$, and $d_x$ the number of those who would die before reaching age $x + 1$; $p_x$ is the proportion of those $l_x$ who would survive to age $x + 1$, and $q_x$ the proportion who would not. Hence:

$$d_x = l_x - l_{x+1}$$
$$q_x = \frac{d_x}{l_x}$$
$$p_x = 1 - q_x = \frac{l_{x+1}}{l_x}$$

The last column shows the *expectation of life*, denoted by $\overset{\circ}{e}_x$, at any age $x$. This is the average number of years that people aged $x$ years may be expected to live, at current rates of mortality. Females have lower mortality rates than males and, therefore, greater expectation of life at all ages.

**Table 28**    Extract from English Life Table No 12, 1960–1962 (males)

| Age | $l_x$ | $d_x$ | $p_x$ | $q_x$ | $\overset{\circ}{e}_x$ |
|---|---|---|---|---|---|
| 0 | 100,000 | 2,449 | 0·97551 | 0·02449 | 68·09 |
| 1 | 97,551 | 153 | 0·99843 | 0·00157 | 68·80 |
| 2 | 97,398 | 96 | 0·99901 | 0·00099 | 67·90 |
| 3 | 97,302 | 67 | 0·99931 | 0·00069 | 66·97 |
| 4 | 97,235 | 60 | 0·99938 | 0·00062 | 66·02 |
| . | . | . | . | . | . |
| 65 | 68,490 | 2,499 | 0·96352 | 0·03648 | 11·95 |
| 66 | 65,991 | 2,625 | 0·96022 | 0·03978 | 11·39 |
| 67 | 63,366 | 2,745 | 0·95668 | 0·04332 | 10·84 |
| 68 | 60,621 | 2,856 | 0·95288 | 0·04712 | 10·31 |
| 69 | 57,765 | 2,959 | 0·94878 | 0·05122 | 9·79 |
| . | . | . | . | . | . |

The student will find it profitable to consider the following questions:

1. If a man's expectation of life at 30 is 40 years, why is his expectation at 40 more than 30 years?
2. Why might the expectation of life of a child one year old be greater than that of a new-born baby?
3. Why would you expect the average age at death of judges or bishops to be higher than that of, say, journalists or shopkeepers?

Deaths are extensively analysed by cause, occupation and social class, as in the Registrar General's Supplement on Occupational Mortality. All these factors raise difficulties of classification. Changes over time in death rates from certain diseases, for example, may be due partly to changes in classification and improvements in diagnosis. The classification generally followed now, for both mortality and sickness, is the International Statistical Classification of Diseases, Injuries and Causes of Death, issued by the World Health Organization. Mortality rates for occupations are less reliable than for causes, since it is necessary to rely on the description given by the person registering the death. Moreover, the occupation stated may not be the one that the deceased followed for most of his life, and which may even have caused his death. Thus, a factory worker who has to leave the factory on account of his health and spends the last year or two of his working life as a car-park attendant will probably be recorded as following the latter occupation.

## Standardized rates

It will be obvious from Table 28 that the crude death rate, like the birth rate, depends greatly on the age distribution of the population, so it may be misleading to compare the death rates of two towns, countries, periods or occupations that differ in their age composition. In 1973, for example, Harlow, a new industrial town, had a death rate of 6·0 per 1000, whereas Worthing, a seaside resort full of retired people, had a rate of 25·3 per 1000.

This difficulty can be largely overcome by the use of *standardized death rates*. The local population is broken up into a reasonable number of age-groups, and a mortality rate calculated for each group. These rates are then applied to a standard population, e.g. the population of the country as a whole. Example 19.1 is purposely oversimplified, but it will illustrate the method.

*Example* 19.1

Calculate a standardized death rate for a town, given the following table:

| Age-group | Town population (thousands) | Deaths per 1000 | Actual deaths | Standard population (millions) |
|---|---|---|---|---|
| 0– 4 | 5 | 10 | 50 | 4 |
| 5–19 | 10 | 1 | 10 | 12 |
| 20–39 | 15 | 2 | 30 | 12 |
| 40–59 | 20 | 6 | 120 | 14 |
| 60–74 | 6 | 25 | 150 | 6 |
| 75 and over | 4 | 120 | 480 | 2 |
| Total | 60 | 14 | 840 | 50 |

The crude death rate is 14 per 1000. The standardized death rate is:

$$\{(10 \times 4) + (1 \times 12) + (2 \times 12) + (6 \times 14) + (25 \times 6)$$
$$+ (120 \times 2)\} \div 50 = 550 \div 50$$
$$= 11 \text{ per } 1000$$

In practice, the population would be broken down into much finer age-groups and standardized on sex as well as age. The standardized rate is, in effect, a weighted mean of age-specific rates, with the standard population as the weighting system.

The above method, called the Direct Method, is impracticable if the age-specific rates of the towns or districts are unknown, and we then fall back on the somewhat complicated Indirect Method. We first calculate the *Area Comparability Factor* (ACF for short) by dividing the death rate of the standard population by the *Index Death Rate*, the death rate which the town would have experienced if it had had the age-specific rates of the standard population. The crude death rate of the town is then multiplied by the ACF to give the indirect standardized rate.

*Example* 19.2

Calculate an indirect standardized death rate for the town of Example 19.1, given that the age-specific rates for the six age-groups in the standard population are 6, 0·5, 1, 7, 35 and 150 per 1000 respectively.

The calculations are best set out by means of a table:

| Age-group | Town population (thousands) | Standard population (millions) | Rates per 1000 of standard population | Deaths at standard rates | |
|---|---|---|---|---|---|
| | | | | In town | In standard population (thousands) |
| 0– 4 | 5 | 4 | 6 | 30 | 24 |
| 5–19 | 10 | 12 | 0·5 | 5 | 6 |
| 20–39 | 15 | 12 | 1 | 15 | 12 |
| 40–59 | 20 | 14 | 7 | 140 | 98 |
| 60–74 | 6 | 6 | 35 | 210 | 210 |
| 75 and over | 4 | 2 | 150 | 600 | 300 |
| *Total* | 60 | 50 | | 1000 | 650 |

Death rate of standard population $= \dfrac{650}{50} = 13$ per 1000

Death rate of town at standard rates $= \dfrac{1000}{60} = 16 \cdot 7$ per 1000

$\therefore \text{ACF} = \dfrac{13}{16 \cdot 7} = 0 \cdot 78$

From Example 19.1, the crude death rate of the town is 14 per 1000,

$\therefore$ indirect standardized rate $= 14 \times 0 \cdot 78$

$= 10 \cdot 92$ per 1000

Another summary measure widely used is the *Standardized Mortality Ratio* (SMR for short), although it is generally given as a percentage. It is simply the actual number of deaths in the town or group expressed as a percentage of the 'expected' number that it would have had at the age-specific rates of the standard population. With the data of Examples 19.1 and 19.2, the SMR of the town would be:

$$\dfrac{840}{1000} \times 100 = 84$$

A low SMR means that the mortality of the town compares favourably with the standard population, allowing for the difference in age (and sex) structure.

All these summary figures, while useful, have their limitations and tell only part of the story.

## Migration

This is probably the weakest point in our population statistics, particularly internal migration from one part of the country to another. Some countries have a system of compulsory registration of such movements, but in Great Britain only aliens are compelled to register. Our migration statistics are derived mainly from the International Passenger Survey. This is a sample survey of passengers entering or leaving the country on all the principal sea and air routes. The sampling fraction varies according to the route and the time of year. Those who say they intend to stay in (or out of) the country for at least a year are recorded as migrants. The grossed-up figures are subject to fairly large sampling errors.

In the 1971 Census, as in the Sample Census of 1966, each person was asked to state his address one year and five years before if he had changed his address. This enabled movements into and within the country to be linked with factors such as age, occupation, industry and socio-economic group.

## Population projections

Estimates of the population between censuses are made by taking the numbers recorded at the last census and making adjustments for births and deaths registered since that date, and for migration. Owing mainly to the lack of any comprehensive system of recording migration, as noted above, such estimates are by no means precise, but they are accurate enough for most purposes.

For the planning of social insurance and other matters it is necessary to estimate future changes in population. Every year, the Government Actuary's Department, in consultation with the OPCS, makes projections for the next forty years, by sex and age. It is emphasized that these figures are not predictions, but estimates based on the best assumptions that can be made at the time about future fertility, mortality and migration. They are published in an annual booklet entitled *Population Projections*.

## Sickness

The chief difficulty about sickness statistics is that sickness as such cannot be recorded and measured. One can only record attendances at the hospital or surgery, absence from work or claims for benefit. Again, much depends on the doctor's diagnosis or willingness to sign a medical certificate. Certain infectious diseases, however, have to be notified, and these are well documented.

After the Second World War, the General Register Office carried out a number of experiments in the collection of morbidity statistics. These included the establishment of a permanent and continuous register of cancer patients (a measure of prevalence and a means of ascertaining progress in the effectiveness of surgery and radio-therapy), a register of mental patients (for long-term follow-up), an analysis of hospital in-patient records (as a measure of sickness treated in hospital), and an analysis of general-practitioner records (as a measure of sickness treated in general practice). The results of these statistical arrangements have been published either as Supplements to the *Annual Statistical Review* of the Registrar General or as special reports in the series *Studies of Population and Medical Subjects*.

Statistics of claims for sickness and injury benefits and of insured persons absent from work are published by the Department of Health and Social Security. The Department make extensive use of sampling methods in assessing and forecasting the numbers of persons claiming various benefits. Besides maintaining a 2 per cent sample (about 500,000) of the insurance records of the insured population, they take samples (of various sizes) of all family allowances, unemployment benefit claims, retirement pensions, and claims for sickness, injury and disablement benefits.

## Publications

The 1971 Census was followed fairly quickly by a *Preliminary Report*, *Advance Analyses* for Great Britain and for counties, and *Summary Tables*, the last being based on a 1 per cent sample. Over the next few years there appeared *County Reports* (three volumes for each county in England and Wales, but only one for each Scottish county) and a large number of detailed reports on topics such as Fertility, Housing, Migration and Qualified Manpower.

Until recently the GRO published an *Annual Statistical Review for England and Wales*, a *Quarterly Return of Births, Deaths and Marriages*, and a *Weekly Return of Births and Deaths*. These have been replaced by a series of OPCS Monitors, a quarterly journal called *Population Trends*, and a number of annual publications on specific subjects. At the time of writing, however, Scotland and Northern Ireland continue to publish their *Annual* and *Quarterly Reports*.

Two annual publications, *Social Trends* and the *Report on the*

*General Household Survey*, contain a great deal of information about the population and such topics as housing, health, education and leisure.

## Suggestions for further reading

*Reports on the Census of Population of England and Wales*
OPCS Monitors
*General Household Survey* (A)
*Population Trends* (Q)
*Social Trends* (A)
Benjamin, B. *Health and Vital Statistics*. London: Allen & Unwin.
Benjamin, B. *Demographic Analysis*. London: Allen & Unwin.
Cox, P. R. *Demography*. Cambridge University Press.

## Exercises

19.1 (*a*) Define a 'standardized rate/average' and explain its purpose.

(*b*) For the following data calculate (i) the crude death rate for the town, and (ii) a standardized death rate.

| Age group | Number of deaths | Town population (thousands) | National population (millions) |
|-----------|------------------|-----------------------------|--------------------------------|
| 0– 5 | 25 | 2·5 | 2 |
| 6–20 | 5 | 5 | 6 |
| 21–40 | 15 | 7·5 | 6 |
| 41–60 | 60 | 10 | 7 |
| 61–75 | 75 | 3 | 3 |
| 76 and over | 240 | 2 | 1 |

(*c*) If it is desired to compare the average prices of consumer goods in two countries by means of standardized averages, explain why the choice of standardizing weights is critical to the results.

(*IoS*)

19.2 By means of age-standardized rates, compare the absence (from all causes) of two categories of employees which, for convenience, may be called juniors and seniors, using the data in the following table:

| Age | Juniors | | Seniors | |
|---|---|---|---|---|
| | Number employed | Percentage lost time | Number employed | Percentage lost time |
| Under 20 | 150 | 2·4 | — | — |
| 20–29 | 350 | 1·6 | 20 | 1·5 |
| 30–39 | 300 | 2·1 | 70 | 3·0 |
| 40–49 | 150 | 2·8 | 180 | 3·5 |
| 50–59 | 50 | 4·6 | 240 | 5·0 |
| 60 and over | — | — | 90 | 7·0 |
| Total | 1000 | 2·2 | 600 | 4·5 |

Comment on your choice of standard population. *(IoS)*

19.3 The following table shows particulars of sickness-absence during a recent year for conductors employed by a bus company, compared with what would be 'expected' based upon the experience of the conductors employed by all bus operators.

| Age group | Average number conductors employed | Actual sickness absence (days) | 'Expected' (rate per conductor) (days) |
|---|---|---|---|
| Up to 29 years | 50 | 240 | 6·0 |
| 30 to 39 years | 53 | 325 | 6·1 |
| 40 to 49 years | 64 | 515 | 8·1 |
| 50 to 59 years | 59 | 1190 | 17·8 |
| 60 and over | 24 | 630 | 23·2 |
| Total | 250 | 2900 | |

Calculate the overall sickness-absence rates using both the Actual as well as the 'Expected' occurrence of sickness. Explain any difference between these two rates in terms of the differences in sickness-absence for the separate age-groups. *(IoT)*

19.4 Demonstrate the calculation of crude and standardized death rates using the following data:

| Age group | Population (in thousands) | Number of deaths | Age distribution of standard population |
|---|---|---|---|
| 0– 9 | 21 | 350 | 221 |
| 10–24 | 30 | 102 | 298 |
| 25–44 | 37 | 229 | 285 |
| 45–64 | 17 | 354 | 149 |
| 65 and over | 5 | 415 | 47 |
| Total | 110 | | 1000 |

*(IoS)*

19.5   Calculate the standardized mortality ratio for woodworkers from the data given below:

| Age group | Deaths registered in the years 1959–63 | 1961 Census population of woodworkers | All males mean annual death rate per 100,000 |
|---|---|---|---|
| 15–24 | 464 | 91,880 | 103 |
| 25–34 | 402 | 88,860 | 112 |
| 35–44 | 805 | 70,910 | 241 |
| 45–54 | 2638 | 80,520 | 707 |
| 55–64 | 5017 | 51,280 | 2171 |

*(IoS)*

19.6   An abridged life table for a particular town gives the following estimates of $l_x$, the number of males surviving to age $x$ out of 1000 live male births:

| Age x years | $l_x$ |
|---|---|
| 0 | 1000 |
| 1 | 976 |
| 5 | 971 |
| 15 | 968 |
| 45 | 918 |
| 65 | 685 |

Each boy starts school on reaching 5 years of age, leaves school on his fifteenth birthday to take a job and retires at age 65. Assuming that the number of male live births in the town is constant at 700 per annum, that mortality is unchanged from that of the above life table and that migration can be ignored, how many school places for boys are needed in the town and how many jobs are required to provide full employment for males? Of those males who start work, what proportion live to draw the retirement pension? *(IoS)*

19.7 The following is part of English Life Table No 12 for females:

| Age x | $l_x$ | $d_x$ | $p_x$ | $q_x$ | $\overset{\circ}{e}_x$ |
|-------|-------|-------|-------|-------|------|
| 20 | 97336 | * | 0·99956 | * | 55·95 |
| 21 | * | * | * | 0·00045 | 54·98 |
| 22 | * | * | * | 0·00047 | 54·00 |
| 23 | * | 48 | * | * | 53·03 |
| 24 | * | 50 | * | * | 52·06 |

Explain the meanings of the column headings and write down their algebraic relationship. Calculate the values marked with an asterisk.

*(IoS)*

# 20 Prices

For detailed prices it is usually necessary to refer to technical, financial or trade journals. Annual average wholesale prices of many articles over a period of years are given in the *Annual Abstract of Statistics*; they are also given each February in *Trade and Industry*. Prices for many varieties of crops, fruit and vegetables, livestock and other agricultural products are given in *Agricultural Statistics, England and Wales, Part II*, published annually. Most information about prices, however, is published in the form of index numbers, and the rest of this chapter will be devoted to an account of the most important ones used in this country, taking wholesale prices first and retail prices afterwards. Import and export price indices will be deferred to Chapter 24.

## Wholesale price index numbers

For a detailed account of the old Board of Trade Wholesale Price Index, the reader should refer to earlier editions of this book. It was designed to measure changes in the value of money relatively to other things, and was derived from 200 price series for all kinds of commodities, including basic materials, intermediate products and manufactured articles. The 'All Items' index was the geometric mean of the price relatives of these 200 items.

In 1951 this index was superseded by the present system, which is based on 'sectors' of the economy or groups of products. The attempt to compress all price changes into a single index has been abandoned, on the grounds that a family of index numbers which are meaningful to industry is more useful than a single index of somewhat doubtful significance. The geometric mean has also been replaced by the more readily understood arithmetic mean.

There are six tables, published monthly in *Trade and Industry*, with the following titles:

1. Price Index Numbers of Materials Purchased.

1a. Price Index Numbers of Materials and Fuel Purchased by Selected Sectors of Industry.

2. Price Index Numbers of Output (home sales).

3a. Commodities Produced in the United Kingdom.

3b. Commodities Wholly or Mainly Imported into the United Kingdom.

5. United Nations Index Numbers of World Export Prices.

The first three relate to materials used or produced by large industrial groups; the next two are more detailed tables for particular commodities. The last table shows figures for four very broad groups of commodities. For some obscure reason there is no Table 4.

Prices are collected for over 10,000 materials and products, mostly from trade associations and large firms. All prices are carefully defined—grade, size or quality, terms of contract etc. They include import and excise duties, but not VAT. The index for any individual commodity is its current home-market price expressed as a percentage of its average home-market price in the base year, at present 1970. Where an index relates to more than one product (which it usually does), it is a weighted arithmetic mean of price relatives, the weights being roughly proportional to the total values of the materials sold or purchased in 1970. Indices for the larger groups are calculated with 'net' weights, i.e. they exclude sales from one firm to another in the same sector to avoid duplication.

The Wholesale Price Indices are widely used in business contracts, in forecasting future costs, and as factors for reducing current values to base-year prices.

### Index of retail prices

The history of this index goes back to the First World War, when the old Cost of Living Index Number was compiled in order 'to measure the percentage increase month by month in the cost of maintaining unchanged the standard of living prevailing among working-class households in July 1914'. Price quotations were collected from retailers, local authorities etc., much as they are now, and each quotation was converted into a price relative based on July 1914. Indices for the five main groups and for All Items were compiled from weighted arithmetic means of the price relatives, the groups and their weights being as follows:

| Group | Weight |
|---|---|
| Food | 60 |
| Rent | 16 |
| Clothing | 12 |
| Fuel and light | 8 |
| Other items | 4 |
| *Total* | 100 |

This weighting was broadly based on the average expenditure shown by 1944 working-class household budgets in a survey conducted by the Board of Trade as far back as 1904. Only the bare necessities of life were taken into account.

As standards of living improved, the Cost of Living Index became seriously deficient. In 1937–8, a new family budget inquiry was held, but the war prevented the results being published and utilized. For the next few years, the prices of food, rent and other basic necessities were severely controlled and, since they were heavily weighted in the index, it showed a much smaller increase in the cost of living than was really the case. It was completely discredited by June 1947, when it was superseded by the present index.

The Index of Retail Prices is compiled by the Department of Employment and published monthly in the *Department of Employment Gazette* and the *Monthly Digest of Statistics*. The change of title is, as we might say in another connection, highly significant. Whereas the old Cost of Living Index had only covered the necessities of life, the present index covers all normal items of expenditure on retail goods and services. Luxuries and 'conventional necessities' have acquired greater importance, not only because of the greatly improved standard of living of the working classes, but also because the coverage of the index has been extended to most small and medium salary earners (see below). Nevertheless, the Index of Retail Prices is still generally referred to in the Press and elsewhere as the 'Cost of Living Index'.

The Household Expenditure Inquiry held in 1953 provided a picture of the average pattern of expenditure of nearly all wage and salary earners, which formed the basis for a sound weighting system for the index. As in the 1937–8 inquiry, however, substantial adjustments had to be made for certain items, such as drink and tobacco, on which people spend more than they care to admit. Fortunately, reliable estimates for these items are available from Customs and Excise.

A continuous inquiry called the Family Expenditure Survey (FES for short), has been running since January 1957. Each year a new sample of about 10,000 addresses is selected, and the households living there are asked to keep detailed records of their expenditure for a fortnight and to supply information about certain regular payments over a longer period, and about incomes.

The FES also collects much information about the characteristics of the households which is of interest to economists, social workers etc., but its main object is to provide a weighting system for the Index of Retail Prices. For this purpose, 'high income' households and pensioner households in which at least 75 per cent of their income derives from state benefits are excluded. The definition of 'high income' naturally varies from year to year.

The weighting of the index is now revised every year in accordance with consumption as ascertained from the FES for the year ended the previous June, all items being revalued at the prices ruling at the date of revision.

Although the FES provides information on all kinds of payments, the index naturally excludes such items as income tax, insurance premiums (other than house insurance) and National Insurance contributions, savings, subscriptions of all kinds and all forms of betting. Capital payments and the repayment part of mortgage payments for house purchase are excluded as being investment rather than consumer expenditure, but the weight for housing includes estimated mortgage interest net after tax relief.

At the time of writing the main groups and their weights, which are derived from expenditure during the period July 1974 to June 1975 adjusted to the prices ruling in January 1976, are as follows:

| | |
|---|---:|
| Food | 228 |
| Alcoholic drink | 81 |
| Tobacco | 46 |
| Housing | 112 |
| Fuel and light | 56 |
| Durable household goods | 75 |
| Clothing and footwear | 84 |
| Transport and vehicles | 140 |
| Miscellaneous goods | 74 |
| Services | 57 |
| Meals bought and consumed outside the home | 47 |
| All items | 1000 |

Prices are collected in various ways. Food prices are collected from retailers by local officers of the Department of Employment in 200 selected areas. Prices of beer and tobacco are obtained from the manufacturers, rent and rates from local authorities, transport from British Rail, London Transport etc., and so on. Every effort is made to ensure comparability and allow for changes in quality, but items like furniture and second-hand cars present obvious difficulties.

Clearly it is necessary to take a relatively few prices as representing a whole group. For example, nine different kinds of furniture are listed and the price index for furniture is the average of the price relatives of these nine 'indicators'. The weight assigned to furniture is, however, based on total expenditure by households on all kinds of furniture.

Separate quarterly indices are published for one-person and two-person pensioner households, defined as households where at least 75 per cent of the total income consists of National Insurance pensions, public assistance etc. They exclude housing, and the weights are based on the expenditure of pensioner households over a three-year period. The indices are slightly higher than the general index.

## Index numbers of agricultural prices

Price indices are compiled by the Ministry of Agriculture, Fisheries and Food for agricultural products (crops, livestock, fruit and vegetables etc.) and materials (fertilizers, feeding-stuffs etc.). For subsidized materials such as fertilizers and lime, the prices used are market prices less subsidy, i.e. the final cost to the farmer. The present base is the period from July 1964 to June 1967 inclusive.

## Consumers' expenditure deflator

The Blue Book on National Income and Expenditure contains two tables of consumers' expenditure: (i) at current prices; (ii) at the prices ruling in a particular year, at present 1970. From these figures a system of price indices can be derived, more broadly based than the Index of Retail Prices, since it covers all consumers and not merely those in a certain income group. The *consumers' expenditure deflator*, as it is called, is, in fact, shown in another table. It differs from the Index of Retail Prices in one important respect, because it relates to current expenditure, i.e. it uses current weighting, while the Index of Retail Prices uses base-period weighting. This is less important now that the latter is revised every year.

## Comparison with pre-war years

It is often required to compare the present value of the pound with its pre-war value, or to construct a continuous series of retail price indices over a long period. It may, therefore, be useful to construct such a series for the period 1914–76. It is necessary to divide this period into three, linking up the component series by multiplication:

(i) For 1914–38 the only available series is the old Cost of Living Index Number. Over this period it is probably fairly reliable. It is necessary to take July 1914 as the average for the year.

(ii) For 1938–47 it is better to use the index numbers given or implied in the White Papers (as they then were) on National Income and Expenditure for 1946 and 1947.

(iii) For 1947 onwards the official Index of Retail Prices may be used, taking the index for 1947 as 100, although 17 June was the actual base date.

Strictly speaking, this is not one continuous series but the product of several.

**Table 29**   UK Index of Retail Prices, 1914–76 (1938 = 100)

| Year | Index | Year | Index | Year | Index |
|------|-------|------|-------|------|-------|
| 1914 | 64  | 1934 | 90  | 1954 | 240 |
| 1915 | 79  | 1935 | 92  | 1955 | 250 |
| 1916 | 94  | 1936 | 94  | 1956 | 263 |
| 1917 | 113 | 1937 | 99  | 1957 | 273 |
| 1918 | 130 | 1938 | 100 | 1958 | 281 |
| 1919 | 138 | 1939 | 102 | 1959 | 282 |
| 1920 | 160 | 1940 | 119 | 1960 | 285 |
| 1921 | 145 | 1941 | 130 | 1961 | 295 |
| 1922 | 117 | 1942 | 139 | 1962 | 308 |
| 1923 | 112 | 1943 | 143 | 1963 | 314 |
| 1924 | 112 | 1944 | 146 | 1964 | 324 |
| 1925 | 113 | 1945 | 149 | 1965 | 339 |
| 1926 | 110 | 1946 | 153 | 1966 | 353 |
| 1927 | 107 | 1947 | 168 | 1967 | 362 |
| 1928 | 106 | 1948 | 181 | 1968 | 379 |
| 1929 | 105 | 1949 | 186 | 1969 | 399 |
| 1930 | 101 | 1950 | 192 | 1970 | 425 |
| 1931 | 94  | 1951 | 201 | 1971 | 465 |
| 1932 | 92  | 1952 | 228 | 1972 | 498 |
| 1933 | 90  | 1953 | 235 | 1973 | 543 |
|      |     |      |     | 1974 | 630 |
|      |     |      |     | 1975 | 783 |
|      |     |      |     | 1976 | 913 |

The complete series, compiled in this way, is given in Table 29 with 1938 as base year.

## Suggestions for further reading

*Monthly Digest of Statistics*
*Economic Trends* (M)
*Trade and Industry* (W)
*Department of Employment Gazette* (M)
*Reports on the Family Expenditure Survey* (A)
*Method of Construction and Calculation of the Index of Retail Prices*

# 21 Labour statistics

## Published sources

The chief source of British labour statistics is the monthly *Department of Employment Gazette* (hereafter abbreviated to 'DE Gazette'). Not only does it contain many pages of regular statistical series on employment, unemployment, earnings, wage rates and other topics, but it frequently has detailed reports on surveys and inquiries, and articles on actual or impending changes in labour statistics.

There is also a *British Labour Statistics Yearbook* that brings together all the main statistics relating to the year in question, usually with data for previous years. A *Historical Abstract* covers the period 1886–1968. Summary tables on various labour topics appear in the *Annual Abstract of Statistics*, the *Monthly Digest of Statistics* and *Economic Trends*. Other publications that frequently contain articles on these subjects are *Social Trends* and the annual reports on the *Family Expenditure Survey* and the *General Household Survey*.

## Employment

From 1948 to 1970 the chief source of employment statistics was the annual count of National Insurance cards every June. In 1970, however, it was decided to abolish these cards and to hold an annual Census of Employment to fill the gap. Inland Revenue supplied their list of 'pay-points', from which employers send PAYE payments, to the Department of Employment, and each June a census form is sent to each pay-point, asking for numbers of employees, males and females, full-time and part-time (less than 30 hours a week) being shown separately. A full census is held every three years, but in intervening years forms are sent only to employers who had three or more employees at the last full census.

The census is supplemented by monthly or quarterly sample

inquiries of employers, the data from which are used to calculate ratios of change since the last census results. Thus the annual census provides the 'benchmark' from which the monthly and quarterly estimates are extrapolated. They are provisional until the results of the next census are available.

The numbers of employers and self-employed are based on the last Census of Population, and up-dated annually by means of estimates of changes derived from a sample of Class II National Insurance cards. Quarterly figures are obtained by interpolation.

The *DE Gazette* gives, once a quarter, a broad industrial analysis of numbers of employees in employment for each Standard Region. It also gives quarterly estimates of the *working population*. This is not, as one might think, the number at work, but rather what an actuary would call the number exposed to risk of work. The working population of Great Britain in June 1976, for example, was made up as follows:

|  | 000's |
|---|---:|
| Employees in employment | 21,997 |
| Employers and self-employed | 1,864 |
| HM Forces | 336 |
| Unemployed | 1,278 |
| Total working population | 25,475 |

(Source: *DE Gazette*, Sept. 1976)

The warnings given earlier about breaks in continuity and the need for careful definition apply particularly to employment statistics. The worst break occurred in 1948, when the National Insurance scheme was introduced. In the old series, which was based on numbers insured against unemployment under the Unemployment Insurance Acts, the total in civil employment at mid-1948 was 19,064,000. In the new and more comprehensive series it was 21,569,000. The increase was due mainly to:

(*a*) the inclusion of about half a million indoor domestic servants and nearly a million workpeople over pensionable age (65 for men, 60 for women), both of which classes were previously excluded; and

(*b*) the counting of about 800,000 part-time workers as full units instead of half units.

Another break followed the revision of the Standard Industrial Classification in 1958, although the Ministry of Labour (as it then was) provided a good link by showing the composition of the working population for several months of 1959 on both old and new classifications. The most important changes were:

(i) The transfer of certain trades (particularly motor repairing and garages, and boot and shoe repairing) from Manufacturing Industries to Services.

(ii) The regrouping of industries, particularly in the Engineering and allied industries.

A further revision of the SIC in 1968 caused only minor changes.

## Unemployment

Unemployment statistics are derived from monthly returns rendered by Employment Offices and Youth Employment Service Careers Offices, showing particulars of the unemployed persons on their registers on a given day in each month. They include not only persons who have fallen out of employment, but school-leavers and others registering for their first employment, so there is usually a seasonal increase in the number registered as unemployed at the end of the school summer term. Persons working on short time or temporarily stood off by their employers who merely register to claim benefit, adult students seeking vacation jobs, and severely disabled persons are all excluded from the statistics.

Tables published in the *DE Gazette* include monthly industrial and regional analyses and quarterly analyses by occupation and by duration of unemployment.

Since unemployment is a vital economic indicator—and sometimes political dynamite—it is important that the significance of changes in numbers unemployed should be correctly assessed. At certain times of the year, e.g. just after Christmas and at the end of school terms, there are strong seasonal effects. Many of the figures are therefore given both actual and seasonally adjusted.

## Labour turnover

Each quarter the *DE Gazette* publishes a table of labour turnover rates in the principal manufacturing industries. The basic data are obtained from returns made by a sample of employers. They relate to periods of four weeks, and include administrative and clerical employees as well as operatives. Persons who are engaged during the period but leave before the end of it are not included; this is not unimportant, for a large number leave within the first few weeks. Losses are calculated from the numbers employed at the beginning and end of the period and the number of engagements, and therefore tend to be understated to the same extent as engagements. On the other hand, losses include persons who have merely changed their employers within the same industry, and the percentage of losses therefore exceeds the total wastage from the industry as a whole.

## Wage rates and earnings

Most manual workers are paid either time rates, a given amount for a unit of time such as an hour or a standard working week with bonuses or higher rates for overtime, or piece rates, a given amount for each unit of work performed. An annual volume entitled *Time Rates of Wages and Hours of Work* sets out for the more important industries the minimum or standard time rates of wages as fixed either by voluntary agreements between organizations of employers and workpeople or by statutory orders. In general, however, wage rates are less important than earnings. 'Earnings' are the worker's pay-packet, his actual pay for the week or other period, including bonuses and overtime, but before deducting income tax, insurance contributions etc.

In comparing wage rates or earnings in different industries, or for different groups of workers, it must be remembered that cash earnings are often supplemented by some form of payment in kind or 'concealed emoluments', not to mention anything so obvious as tips. Clerical staff may receive luncheon vouchers, railway employees are allowed free or cheap travel, and so on; some pension schemes are contributory and others are non-contributory. Comparisons should also take into account the proportions of men and women, skilled and unskilled workers etc., in the groups or industries concerned.

## Index numbers of wage rates and hours worked

Indices of basic weekly and hourly wage rates and hours of work appear monthly in the *DE Gazette* and the *Monthly Digest of Statistics*. In the Index of Basic Weekly Rates of Wages, the figures for All Industries and for industry groups are weighted arithmetic means of the indices for their component trades, the weights being approximately proportional to the total wage bills of those trades in 1970.

Indices of 'normal' weekly hours are derived from the information received about changes in the basic working week. These indices are then divided into the indices of weekly wage rates to obtain indices of hourly rates. Thus, taking the figures for 1975 shown in Table 30,

$$175 \cdot 4 \div \frac{99 \cdot 6}{100} = 176 \cdot 2$$

Table 30 shows the three series for men (All Industries and Services) for a run of years. The last column shows the continuing effect of the movement for a shorter basic working week, which began in mid-1959 and has generally resulted in the basic week being reduced by four hours. The average number of hours worked has not been reduced by this amount: it simply means that more time has been counted as overtime.

**Table 30** Indices of rates of wages and normal weekly hours for men—all industries and services 1972–5 (31 July 1972 = 100)

| Year | Basic weekly rates of wages | Basic hourly rates of wages | Normal weekly hours |
|------|------|------|------|
| 1972 | 101·5 | 101·5 | 99·9 |
| 1973 | 114·9 | 115·2 | 99·8 |
| 1974 | 136·4 | 136·9 | 99·6 |
| 1975 | 175·4 | 176·2 | 99·6 |

## Earnings and hours of employees

The two principal sources of information on earnings and hours worked are (*a*) an annual inquiry held each October, covering manual workers in manufacturing and various other industries, (*b*) the New Earnings Survey held every April, based on a 1 per cent sample of employees. There are other minor inquiries, but we shall confine ourselves to these two.

At the time of writing, the October inquiry, which is still voluntary (unlike other inquiries by the D of E), covers about five million workers in about 37,500 establishments. The principal employments excluded are agriculture, coal-mining, railways, shipping, distribution and catering trades, commerce and banking, entertainment and domestic service. Statistics of earnings in agriculture, railways and coal-mining, however, are compiled by the relevant bodies, such as the National Coal Board.

The information required is the number of manual workers at work in a specified week, their total wages and total hours worked, distinguishing between men (21 and over), women (18 and over), youths and boys, and girls, also part-time workers—men and women. The results of the inquiry are published in tables showing, for each industry and for each category of workers,

(*a*) Number of wage-earners covered by the returns.
(*b*) Average earnings in the week in question.
(*c*) Average number of hours worked in that week.
(*d*) Average hourly earnings in that week.

Separate figures are shown for full-time and part-time women employees. There are also analyses by industry group and Standard Region together.

The Department of Employment is careful to point out that: 'In view of the wide variations, between different industries, in the proportions of skilled and unskilled workers, in the opportunities for extra earnings from overtime, night-work and payment-by-results schemes and in the amount of time lost by short-time working, absenteeism, sickness etc., the differences in average earnings shown in the tables should not be taken as evidence of, or as a measure of, disparities in the ordinary rates of pay prevailing in different industries for comparable classes of workpeople employed under similar conditions.'

The New Earnings Survey was first held in 1970. Employers are required to give detailed information about employees whose National Insurance numbers end in a specified pair of digits, the make-up of their earnings in a particular pay-period and of their hours worked, and any special factors affecting pay. This enables the Department of Employment to calculate averages, quartiles and distributions of weekly and hourly earnings, the effects of overtime, incentive schemes etc.

## Clerical salaries analysis

The survey with the above title is carried out every year, not by the Department of Employment but by the Institute of Administrative Management. Employers are asked to analyse their clerks by six carefully defined grades and 35 weekly salary ranges. The report consists principally of Area Tables giving, for each region, large town, or other well-defined area, the median and quartile salaries of each grade of clerk. Care is taken to avoid spurious accuracy, the figures being rounded and shown only for areas with more than six establishments employing a total of 50 clerks or more.

## Suggestions for further reading

*Department of Employment Gazette* (M)
*British Labour Statistics Yearbook*
*British Labour Statistics, Historical Abstract 1886–1968*
*New Earnings Survey* (A)
*Clerical Salaries Analysis (The Institute of Administrative Management)* (A)

# 22 Production

## The nature of production statistics

In this chapter we take 'production' to mean the growth, extraction or manufacture of tangible goods, which can be counted or measured, as distinct from services, such as finance, education, transport and retail trade. Production can be measured at various stages: we can record the amount of coal mined at a colliery, the quantities of dyestuffs made from that coal, the quantities of yarn dyed with those dyestuffs and the number of table cloths, curtains etc. made with that yarn. In other words, we can measure output of raw materials, intermediate goods, and finished goods. Often the finished goods of one industry are the raw materials of another. This can, if we are not careful, lead to double counting, but it is often convenient to measure output at different stages of a process, even within the same factory.

A distinction is sometimes made between producers' goods and consumers' goods, but this is not watertight. Agricultural products are consumers' goods when eaten by human beings, but producers' goods when eaten by cattle. Cars may be producers' goods in working hours and consumers' goods during hours of leisure.

For many purposes it is convenient to deal with Industrial Production as a group, i.e. manufacturing industry, mining and quarrying, construction and public utilities, and to omit agriculture or treat it separately. For one thing, it is impossible to measure output of crops and livestock month by month or even quarterly in any meaningful way: the results of a single month's work cannot be estimated and may not even be visible until long afterwards. We shall therefore deal briefly with agriculture, then consider industrial production in more detail.

## Agricultural production

The first Census of Agriculture was taken in 1865, and a full census of all but the smallest holdings is taken every June. The main publication is the annual *Agricultural Statistics: United Kingdom*, which gives the results of the censuses, numbers of livestock, acreage and production of crops etc. There are also separate publications, more detailed, for England and Wales (together), Scotland, and Northern Ireland. Figures of production and stocks of a wide range of agricultural products are given in the *Annual Abstract* and the *Monthly Digest of Statistics*. A large number of other publications containing statistics of agriculture, fisheries etc. are issued in this and other countries, and by such bodies as the Food and Agricultural Organization of the United Nations.

## Statistics of industry: the new system

The reader who is interested in the history of production and distribution statistics should refer to earlier editions of this book; he may well find one in his local library. We shall start now in the late 1960s, by which time a three-tier structure had been developed, comprising a Census of Production or Distribution roughly every five years, less detailed annual surveys in between, and very detailed monthly or quarterly statistics, usually for specific products, obtained from a sample of firms as *indicators* of short-term movements. (The word 'indicator' is commonly used to mean an item or series, often an index number, selected or compiled from short-term statistics as representing some general trend.)

The formation of the BSO heralded a new approach. The very detailed quinquennial Census of Production, like that for 1968, which asked for sales and purchases of individual products and materials, was replaced by much simpler annual censuses, beginning with 1970. The emphasis is now on industries rather than products: the product-based monthly and quarterly inquiries have been superseded by industry-based quarterly inquiries in which each establishment has only one form to complete. Moreover, the forms ask for figures of sales rather than production, which was sometimes difficult to estimate. They are supplemented by monthly inquiries for a few selected industries regarding total sales, stocks, orders on hand etc.

In general, the Census of Production covers undertakings in all fields of industrial production, including gas, electricity and water supply, mines and quarries. The line between production and services, however, is somewhat arbitrary: for example, laundry, cleaning and dyeing trades were at one time included, but are now excluded. Repair work is generally covered by the Census of Distribution, but aircraft, ship and locomotive repairs are regarded as an essential part of industry and are included in the Census of Production.

The Census of Production collects information on:

I Working proprietors
II Employment
III Wages and salaries
IV Stocks
V Capital expenditure
VI Work given out
VII Transport payments and Post Office parcel services
VIII Purchases of materials and fuel
IX Sales and work done

The *gross output* of a trade is defined as the total value of goods made and other work done during the year. The *net output*, which in many ways is more important, is equal to the gross output less the cost of all purchased materials, fuel etc. used, any payments for work given out to other firms or persons, any transport payments to other firms, and any Customs and Excise duty. In short, net output is the value added by manufacture to the raw materials, a concept with which VAT has made many of us familiar.

Both gross and net output, as defined here, exclude any VAT etc. levied on the finished products, and are valued before any deductions for subsidies. In other words they are valued 'at factor cost', i.e. at the prices paid to the various factors of production—labour, management, land, capital etc., including profits, whether distributed or not. The Census Reports drive the point home by saying '. . . it (net output) constitutes the fund from which wages, salaries, insurance, pensions, hire of plant and machinery, payments for repairs and maintenance, costs of operating road vehicles, rents, rates and taxes, advertising and other selling expenses and all other similar charges have to be met, as well as depreciation and profits'.

The results of the Census of Production are published in about 160 annual Business Monitors bearing the prefix PA. Thus, PA 274 is the Census report for Paint. Quarterly Production Monitors (prefix PQ) give detailed sales figures for over 4000 products, often accompanied by figures of imports, exports, prices etc. In addition, statistics of sales, production, stocks etc. of many products are given in the *Annual*

*Abstract* and *Monthly Digest of Statistics, Trade and Industry*, and more specialized publications like *Energy Trends* and *Housing and Construction Statistics*.

## The UK index of industrial production

This index, which has been described in *Statistical News* as 'by far the most important industrial indicator', is published monthly in *Trade and Industry* and in the *Monthly Digest of Statistics*. The present base year is 1970, so that the index for any month gives a comparison with the average month of 1970. Agriculture, trade, transport, finance and all other public or private services are excluded for reasons discussed above. In fact, it covers the same industries as the Census of Production, together with construction, which was included in the old quinquennial censuses but is now omitted from the annual censuses.

Separate indices are published for each of the main industry groups in the Standard Industrial Classification, their weights being roughly proportional to the net output of the corresponding industries in 1970, as revealed by the Census of Production for that year. The index thus gives a comparison of the net output of the current month with that of the average month of 1970, both valued at 1970 'prices', price in this case meaning net output per unit of quantity.

The index is built up from volume relatives of hundreds of individual series, the data being generally derived from the quarterly and monthly inquiries. Whenever possible, they are based on output in tonnes or some other physical measure, but often this is impracticable. Sometimes it is necessary to use sales or deliveries instead of production. Failing that, 'deflated value' may be used, i.e. value of sales or output deflated by a suitable wholesale price index, as in the case of clothing. In a very few cases, the indicator used is numbers employed or a basic raw material, e.g. consumption of newsprint for newspapers and periodicals.

Heterogeneity is a constant problem. As with price indices, an index of production inevitably fails to take into account many changes in quality and variations in size, grade etc. that might well vitiate the index if they were biased in one direction, e.g. if there were a tendency to produce larger motor-cars, better houses or stronger beer. In some cases, such as drugs, where there is infinite variety and new products are constantly appearing, it is impossible to give an estimate of volume of production with any confidence, just as it is impossible to compile a reliable price index for them.

Adjustments are made for variations in the number of working days in the month, so that in effect the indices compare the average weekly rates of production in different months, but they still reflect fluctuations due to public holidays, annual holidays and other seasonal influences. Monthly indices are therefore shown seasonally adjusted as well as unadjusted. The seasonal factors are recalculated each year: they are said to be 'designed for the sole purpose of making the trend more readily discernible'.

## Input–output tables

The relations between one industry and another may be illustrated in the form of tables showing sales by one industry to another and purchases by one industry from another. This is really a two-way analysis of the gross output of each industry group etc., showing the various components of input and output of each group and arriving at the same total in both cases. For completeness, these input–output tables, as they are called, must include sales to final consumers, exports, investment and stock appreciation on the one hand, and labour costs, profits etc. on the other. Table 31 shows a simplified input–output table for a hypothetical country with only three industries.

The output of each industry group as shown in such tables is 'free of duplication', i.e. it excludes all transactions between establishments within that group. It therefore differs from gross output as defined in

**Table 31**  A simplified input–output table

| Sales by industry group | Purchases by industry group | | | | | |
|---|---|---|---|---|---|---|
| | Industry A | Industry B | Industry C | Final buyers | Exports | Total output (£ million) |
| Industry A | — | 25 | 50 | 170 | 45 | 290 |
| Industry B | 40 | — | 90 | 35 | 15 | 180 |
| Industry C | 100 | 20 | — | 60 | 30 | 210 |
| Imports | 20 | 15 | 30 | 35 | — | 100 |
| Taxes on expenditure *less* subsidies | 5 | — | 10 | 15 | — | 30 |
| Wages, salaries etc. | 90 | 80 | 25 | — | — | 195 |
| Gross profits | 35 | 40 | 5 | — | — | 80 |
| *Total input* | 290 | 180 | 210 | 315 | 90 | 1085 |

the Census of Production, which includes sales by establishments to other establishments in the same industry.

Various other tables can be compiled from the original table, such as an analysis of the final output of each industry group into the percentages contributed by each group (including itself) in terms of net output, or a table showing what proportion of each industry's output is exported, directly or indirectly.

Input—output tables for the United Kingdom are now published annually. They are compiled from Census of Production data supplemented by a variety of information for industries outside the scope of the census.

## Output per head

A table with this heading appears in the *Monthly Digest of Statistics* and gives indices, for a few broad groups, of:

Output (identical with the Index of Industrial Production)
Employment
Output per head

The last item is simply the first divided by the second.

It is not proposed to discuss at great length what is primarily not a statistical, but an economic problem. It is necessary, however, to sound a warning about the uncritical use of ratios or indices purporting to measure productivity. The commonest of these is ratio of output, whether in physical or financial terms, to numbers employed or man-hours worked. This is often called 'labour productivity' to distinguish it from other ratios involving capital employed, materials used etc. It is a misleading term, as it suggests that changes in output per man are due to labour alone whereas they may well be due to increased mechanization, improvements in technique and other causes.

This is not to say that it is useless to study variations in output per man or per man-hour. In certain industries, e.g. cotton, some very useful work has been done on these lines, but the farther such calculations get from the factory floor, the more academic they become. Broad indices of output per worker for a whole firm or industry are of little practical value, and may be misleading in such matters as wage negotiations.

International comparisons must also be treated with caution. An industry or product may represent different things in different countries (think of the American 'drug-store'). It would be unreasonable to compare the number of cars produced in the United States with the number of much smaller cars produced in the United Kingdom, and where the output of an industry can be expressed only in terms of money, rates of exchange may present difficulties. It may, however, be easier to compare the rates of change of productivity in different countries.

## Suggestions for further reading

*Annual Abstract of Statistics*
*Monthly Digest of Statistics*
*Trade and Industry*
*Reports on the Census of Production*
*Input—Output Tables for the United Kingdom* (A)
*Studies in Official Statistics*, No 25, 'The Measurement of Changes in Production'
*Business Monitors* (Production series)
*Energy Trends* (M)
*Housing and Construction Statistics* (Q)

# 23 Internal trade

We turn now from production to what might be called the rest of the process, i.e. the distribution and sale of goods to the consumer, including such services as catering, hairdressing, laundries and most repair work. We shall also include in this chapter brief sections on stocks, capital expenditure and transport.

The collection of distribution statistics is still, at the time of writing, undergoing an upheaval similar to that from which production statistics have already recovered. In place of the comprehensive quinquennial Census of Distribution and occasional large-scale inquiries into wholesaling, there will be a system of annual sample inquiries collecting regular information on important matters and including less important topics in rotation. The first such inquiry will relate to 1976.

At the same time there is a wide range of monthly and quarterly inquiries similar to those for production. Much of the resulting information is published in *Trade and Industry* and the Service and Distributive Series of *Business Monitors*.

As the Census of Distribution is still of considerable interest and there will be nothing to replace it for several years, a brief account of it will now be given.

## The census of distribution

The Census of Distribution and Other Services, to give it its full title, provides valuable information for business firms, local authorities, market research workers etc., and forms a framework for the Index Numbers of Retail Sales and other short-term statistics, and for any sample surveys that may be required. It covers all establishments engaged in retailing, hiring out and repair of consumer goods, and related service trades.

The first full Census was taken for 1950, after a pilot sample census for 1947. The second full census was for 1961 and the third for 1971, sample censuses being taken for 1957 and 1966. The 1971 Census asked for figures of turnover, purchases, numbers employed, wages and salaries, stocks, capital expenditure, hire-purchase sales, debts, transport costs and floor space. Certain terms which are used in the Census Reports need to be defined:

(*a*) *Establishment:* a single shop or a single branch of a multiple or other organization.

(*b*) *Persons engaged:* not merely employees, but all who work in the business, including working proprietors, members of the owner's family and part-time workers.

(*c*) *Turnover:* total trading receipts before deducting any outgoings.

(*d*) *Gross margin:* turnover less purchases plus increase (or less decrease) in stocks during the year. Rather like net output, but making no deduction for fuel and other goods, such as display materials, which are used, not sold.

(*e*) *Stockturn:* the ratio of annual turnover to stocks held at the end of the year.

Retail organizations are classified into three groups:

Co-operative societies
Large multiple organizations (with ten or more branches)
Independent organizations

The 1950 Census included the wholesale trade, catering and motor trades, but these are now covered by separate inquiries. There was a large-scale inquiry into the wholesale trade for 1974 and there will be one for catering for 1977, based on a sample from the new Business Register. These also will be superseded in time by annual sample inquiries.

## Index numbers of retail sales

An Index of Retail Sales was compiled for several years, before and during the Second World War, by the Bank of England in collaboration with trade associations. In 1947 a more detailed return was introduced and the work was taken over by the Board of Trade. For some years two series, one for Large Retailers and one for Independent Retailers, appeared monthly in the *Board of Trade Journal* and the *Monthly Digest of Statistics*, but they suffered from an inadequate and possibly biased sample of retailers and from the lack of a proper weighting system.

The 1950 Census of Distribution provided a much-needed sampling frame and a reliable weighting system, and indices of weekly values of sales were published for various kinds of business; the panel of retailers was also extended to cover nearly all retail trade. In 1961, indices of volume were added. These indices are still compiled in much the same way, being revised after each Census of Distribution, and based on the year of that census. Those for broad groups are weighted arithmetic means of the component indices, with weights proportional to total turnover in the census year. Values include VAT (or purchase tax before April 1973).

Table 32 shows some of these indices of volume and value, most of which are seasonally adjusted. The tendency now is to regard seasonally adjusted figures as those of real significance, and to use more sophisticated methods of making these adjustments. There is, however, no satisfactory way of removing the effects of changes or expected changes in VAT, hire-purchase regulations etc., or of abnormally good or bad weather and, however refined the methods may be, the seasonal factors are subject to considerable uncertainty.

## Hire purchase

Statistics of hire purchase are an important indicator of the country's economy. There are two main types of transaction: (i) hire-purchase agreements under which the customer hires the goods but does not own them until he has paid off the credit allowed him; (ii) credit-sale agreements under which the goods become his property from the out-set. The former usually relate to large durable goods like cars and furniture; the latter to clothing, bicycles etc. Credit may be obtained from finance houses, durable goods shops and department stores, or other instalment credit retailers such as mail-order houses.

Various tables published monthly in *Trade and Industry* give figures of new credit extended, repayments and debt outstanding. The more detailed tables of new credit show percentage changes compared with a year earlier. The information is obtained from a sample of finance houses, stores and other sources of credit.

## Stocks and capital expenditure

These two items are closely connected and may be discussed together as two aspects of capital formation. Quarterly returns of stocks and capital expenditure are rendered voluntarily by large numbers of firms in manufacturing industry, distribution and service trades, and the results are summarized in *Trade and Industry* and, in less detail, in the *Monthly Digest of Statistics*. The articles dealing with stocks show changes in the seasonally adjusted volume (i.e. value at base-year prices) held by various industry groups and kinds of distributors and, in less detail, changes in book value of stocks.

The articles on fixed capital expenditure also show base-year values, seasonally adjusted, by various industry groups and types of distributive and service trades. The totals for manufacturing industry and for distributive and service trades (but not for industry groups) are broken down into New Building Work, Vehicles, and Plant and Machinery. Before 1963 these items were shown both at current

**Table 32**   Index numbers of retail sales in Great Britain

(*Monthly Digest of Statistics*, November 1976)
(Sales: weekly average 1971 = 100)

| | | | Non-food shops | | | |
|---|---|---|---|---|---|---|
| | All kinds of business | Food shops | Total | Clothing and footwear shops | Durable goods shops | Other non-food shops |
| | | | Seasonally adjusted | | | |
| Sales in 1971 (£ million) | | | Volume | | | |
| 1973 | 110·7 | 101·0 | 117·7 | 109 | 127 | 118 |
| 1974 | 109·9 | 101·8 | 115·9 | 107 | 123 | 117 |
| 1975 | 107·9 | 99·3 | 114·2 | 108 | 121 | 115 |
| 1975 1st quarter | 111·3 | 101·1 | 118·6 | 111 | 129 | 118 |
| 2nd quarter | 108·9 | 98·6 | 116·3 | 108 | 130 | 116 |
| 3rd quarter | 105·5 | 97·9 | 111·0 | 108 | 111 | 112 |
| 4th quarter | 105·8 | 99·4 | 110·4 | 105 | 112 | 112 |
| 1976 1st quarter | 107·3 | 99·5 | 113·1 | 108 | 118 | 114 |
| 2nd quarter | 106·7 | 99·3 | 112·1 | 106 | 121 | 112 |
| 3rd quarter | 108·7 | 98·8 | 115·8 | 109 | 127 | 115 |
| 1975 July | 105·4 | 97·7 | 111·0 | 110 | 109 | 112 |
| August | 104·8 | 97·5 | 110·2 | 109 | 113 | 110 |
| September | 106·1 | 98·6 | 111·6 | 106 | 112 | 114 |
| 1976 July | 108·3 | 99·5 | 114·7 | 107 | 127 | 114 |
| August | 108·9 | 99·5 | 115·8 | 109 | 128 | 115 |
| September | 108·8 | 97·7 | 116·8 | 111 | 127 | 116 |

## Inland transport

Until fairly recently, transport statistics were compiled by various bodies with apparently little co-ordination. The Department of the Environment published two annual reports, (i) *Highway Statistics*, containing information on the nation's roads and road vehicles, and (ii) *Passenger Transport in Great Britain*, giving details of all kinds of values and at base-year values, not seasonally adjusted, so that it was possible to compile indices of costs of construction, vehicles, and plant and machinery. Although these figures are no longer published, they can be obtained from the Department of Trade and Industry on request.

**Table 32 (contd.)**

| All kinds of business | | Food shops | | | |
|---|---|---|---|---|---|
| | Unadjusted | Total value of sales | Grocers | Butchers | Greengrocers, fruiterers |
| Value | | | | | |
| 15,895 | | 6,693 | 4,062 | 905 | 381 |
| 126·9 | 126·9 | 122·8 | 124 | 127 | 116 |
| 146·7 | 146·7 | 144·5 | 149 | 147 | 133 |
| 174·5 | 174·5 | 174·6 | 180 | 170 | 161 |
| 165·8 | 154 | 158 | 164 | 166 | 137 |
| 173·7 | 168 | 170 | 177 | 161 | 177 |
| 177·0 | 171 | 175 | 182 | 161 | 172 |
| 182·1 | 207 | 197 | 201 | 191 | 160 |
| 190·2 | 176 | 187 | 191 | 186 | 172 |
| 193·9 | 187 | 197 | 202 | 178 | 210 |
| 203·8 | 197 | 203 | 208 | 189 | 186 |
| 175·3 | 171 | 172 | 179 | 153 | 192 |
| 175·8 | 168 | 176 | 184 | 156 | 172 |
| 179·2 | 173 | 176 | 181 | 172 | 157 |
| 199·9 | 195 | 199 | 205 | 175 | 204 |
| 204·3 | 196 | 206 | 214 | 188 | 188 |
| 206·4 | 198 | 203 | 206 | 202 | 171 |

passenger transport and the British Road Federation publishes an annual report entitled *Basic Road Statistics* 'in an attempt to bring some order into the confusion which arises from the use of contradictory or conflicting road statistics'.

However, the position has improved. *Highway Statistics* and *Passenger Transport in Great Britain* have been replaced by an annual and extremely comprehensive *Transport Statistics: Great Britain*, covering all forms of inland surface transport including canals, pipelines and coastal shipping.

## Suggestions for further reading

*Trade and Industry* (W)
*Monthly Digest of Statistics*
*Economic Trends* (M)
*Reports on the Census of Distribution and other Services*
Department of Industry. (1975). Developments in the Statistics of the Distributive Trades. *Statistical News*, No 31.
Makepeace, R. W. (1972). Preparing for the Census of Distribution for 1971. *Statistical News*, No 16.

# 24 Overseas trade and Balance of Payments

Importers and exporters are compelled by law to give particulars of all goods imported into, or exported from, the United Kingdom, including the value and, where required, quantity and the country of consignment (see below). Before 1970, there were different classifications for imports and exports, and importers had to classify their goods in two ways—for duty and for statistical purposes. Now both imports and exports follow the 'Customs and Excise Tariff and Overseas Trade Classification'.

Certain classes of goods are excluded from the detailed import and export statistics, such as personal effects and parcels brought or taken by passengers for private use, ships' stores and bunkers (coal and oil), and atomic energy materials. They are, however, allowed for in the Balance of Payments (see below), and more recently in the summary figures issued each month to the public. Exports are valued fob ('free on board'), which represents the cost to the purchaser abroad, including packing, inland and coastal transport in the United Kingdom and all other expenses up to the point where the goods are deposited on board the exporting vessel or aircraft.

Imports are valued cif ('cost, insurance and freight'), which broadly means the cost to the importer up to the port of entry. It does not include Customs duty or purchase tax. Again, however, the summary figures are adjusted to agree with the Balance of Payments, and the import figures totals are presented fob.

Imports are classified as received from the country of consignment, i.e. the country from which they were originally dispatched to the United Kingdom, not necessarily the country of origin or manufacture. Indeed, the importer may not know the country of origin. Exports are classified as dispatched to the country of final consignment, not necessarily the country of unshipment or consumption.

Before 1970 a distinction was made between 'Exports' and 'Re-exports', but there is now no such distinction.

## Trade returns of the United Kingdom

The original documents are sent to the Statistical Office of Customs and Excise, where the data are tabulated and summarized. The results appear in the following publications:

(*a*) *Annual Statement of the Overseas Trade of the United Kingdom*, published in five volumes. Volume I is a summary volume, giving the total quantity (where appropriate) and value of imports and exports of each item classified. Volumes II and III repeat this information for imports and exports respectively, with details for the more important countries concerned. Volume IV takes every country in turn and gives particulars of the principal goods imported or exported. Volume V gives details of the principal commodities for each port (sea or air) of the United Kingdom.

(*b*) *Overseas Trade Statistics of the United Kingdom*, published monthly, showing imports and exports of each item classified, with some country detail, for the current month and the year to date.

(*c*) *Statistics of Trade through United Kingdom Ports*, published quarterly. It shows total trade in tonnes and value through individual seaports and airports for the last quarter and the year to date, with similar figures for the previous year. There is a broad commodity breakdown.

*Trade and Industry* contains monthly articles on United Kingdom trade, giving summary tables of imports and exports by area and by commodity class (but not both together), all seasonally adjusted; quarterly articles on World Trade and Production, giving summary figures for world trade and for imports and exports of OECD countries, all in US dollars and all seasonally adjusted; also an annual article on World Exports and Manufactures. *Economic Trends* gives annual and quarterly figures, the latter being seasonally adjusted. *Business Monitor M10* (annual), Overseas Trade Analysed in Terms of Industries, analyses, the commodities imported or exported according to the industry whose main products they are. Most quarterly Business Monitors include imports and exports of the relevant commodities.

## Unit value and volume index numbers

For many years the Department of Trade and Industry has published index numbers of volume of imports and exports, sometimes accompanied by indices of import and export prices. The term 'unit value' is now used officially in preference to 'price' because the indices are compiled not from price quotations for specific products, but from average values obtained by dividing the total value of a commodity or heading (which may not be quite homogeneous) by the total quantity.

**Table 33**  Summary of UK Balance of Payments 1975

(Figures in £ million)

| | |
|---|---:|
| Exports (fob) | 18,768 |
| Imports (fob) | 21,972 |
| Visible balance | −3,204 |
| Invisibles (net): | |
| Government services | −646 |
| Government transfers | −379 |
| Private services: | |
| Sea transport | +12 |
| Civil aviation | +110 |
| Travel | +239 |
| Other | +1,334 |
| Private transfers | −88 |
| Interest, profits and dividends | +949 |
| Invisible balance | +1,531 |
| Current balance | −1,673 |
| Investment and other capital flows | +188 |
| Balancing item | +6 |
| Balance for official financing | −1,479 |

The present series of Unit Value Indices was introduced in 1963, together with a new series of Volume Indices. They are published monthly in the same article in *Trade and Industry* and, at the time of writing, are both based on 1970 and compiled with 1970 weights, which means that their product does not necessarily agree with the index of total value. In other words, it is officially recognized that one cannot have at the same time a volume index which reflects only changes in quantities, and a price index which reflects only changes in prices, whose product is equal to the index of total value (see Chapter 17).

In both Unit Value and Volume Indices it is necessary to exclude from the calculations items which are subject to erratic fluctuations in size, quality etc., where a quantity or average-value figure might be misleading, but all commodity headings that appear to comprise reasonably homogeneous products are included.

An index of particular interest to Government is known as the *Terms of Trade*. This is the ratio of the Export Unit Value Index to the Import Unit Value Index, expressed as a percentage. When this is greater than 100, the terms of trade are said to be favourable; when it is less than 100, they are adverse.

## Balance of payments

The balance of payments covers all the economic and financial transactions between residents of the United Kingdom and those of other countries. For many years, these transactions were summarized in three accounts, the Current Account, the Long-Term Capital Account, and Monetary Movements. However, it became increasingly difficult to distinguish between 'long-term capital' and 'monetary movements', and this form of presentation was abandoned in 1970 in favour of a simpler and more intelligible approach.

The emphasis is now on the *current balance*, i.e. the difference between current income and expenditure, rather similar to profit or loss over the year, and the *balance for official financing*, which in theory is equal to the current balance together with *investment and other capital flows*. The balance for official financing is known precisely from official records of currency transactions, but since the other two items are built up from estimates, there is usually a substantial discrepancy called the *balancing item*.

The current balance is derived from *visible trade*, i.e. imports and exports of goods, both valued fob, and *invisibles*, which fall into three main groups:

(*a*) Services, such as shipping and civil aviation, banking and insurance, travel, and Government expenditure abroad on armed forces, embassies etc.

(*b*) Interest, Profits and Dividends, i.e. income from overseas investments. In the case of British firms operating abroad, all profits, not merely distributed profits, are included, since retained profits are regarded as income reinvested. They therefore appear twice, as a credit item in this group and as a debit item in Private Investment.

(*c*) Transfers, such as gifts, legacies and aid to under-developed countries.

Table 33 shows how the current balance for 1975 was arrived at, and how the three major items are reconciled. The balancing item is unusually small and may well be much larger when these figures are revised in later years.

The import and export totals are believed to be fairly accurate, but the balance of visible trade, being the difference between two large and (in normal years) almost equal quantities, is relatively less reliable. The main weaknesses in the invisibles are private transfers and 'other services' (finance, royalties and films, postal services etc.). Figures for civil aviation are based on quarterly returns from airlines, and those for shipping on censuses about every fourth year, sample inquiries in other years and small quarterly sample surveys by the General Council of British Shipping. Estimates for travel are based on interviews of a stratified sample by the International Passenger Survey, and are believed to be fairly reliable.

Detailed tables and notes are given in the annual 'Pink Book' entitled *United Kingdom Balance of Payments*, and quarterly estimates are published in *Economic Trends* with a detailed commentary. Estimates for individual quarters, particularly of the current balance, are less reliable than the annual figures.

The remarks made about reliability here and elsewhere should not be taken as in any way a reflection on the competence of our Government statisticians. They are well aware of the limitations of their statistics and do the best they can with imperfect data.

## Suggestions for further reading

*Annual Statement of the Overseas Trade of the United Kingdom*
*Overseas Trade Statistics of the United Kingdom* (M)
*Trade and Industry* (W)
*Economic Trends* (M)
*United Kingdom Balance of Payments* (A)
Allen, R. G. D. & Ely, J. E. *International Trade Statistics*. London & New York: John Wiley.

# 25 National income and expenditure

## Sources

The three main sources from which the national income statistics are derived are the Census of Production, the statistics of tax assessments, and the accounts of the Central Government. There are a large number of minor sources such as the statistics of retail sales, Customs and Excise statistics, Balance of Payments figures, accounts of local authorities and public corporations, and so on. Since the data have been collected for different purposes and on different bases, it is a formidable task to fit them together, and the result has been aptly described as 'essentially a synthetic product'.

As already mentioned in Chapter 18, quarterly estimates of national income and expenditure are published in *Economic Trends* in the form of a special article. They also appear without commentary in the *Monthly Digest of Statistics*. Quarterly capital and financial accounts for each sector of the economy appear regularly in *Financial Statistics*. Annual figures covering a period of years are given in more detail in a Blue Book on National Income and Expenditure, compiled by the Central Statistical Office; it will be referred to throughout this chapter as 'the Blue Book'. *Regional Statistics* provides figures for Standard Regions.

## Threefold nature of national income

National Income may be defined as a measure of the goods and services becoming available to the nation for consumption or investment. It may be studied from three aspects, viz. how it is produced, how it is distributed, and how it is spent. Now the only practicable way of measuring the value of all goods produced and services rendered is to ascertain the incomes that make up the cost of production, i.e. wages, salaries etc., including company profits, whether distributed or not. National Product and National Income are therefore equal by definition and derived from the same data, although their components may be shown differently in the tables. It must be noted, however, that National Income is generally taken to mean the net national income after allowing for capital consumption (see page 284).

Again, since incomes are either spent or saved, National Income is equal to expenditure on consumption plus addition to wealth, i.e. to National Expenditure defined in such a way as to include investment. It can therefore be built up from estimates of expenditure and investment, but this time the data are different and the resulting figure of National Expenditure will probably differ from that of National Income or Product, owing to errors of estimation.

Whichever approach is used, the aggregate itself is of limited interest. To economists, businessmen and Government, the individual items that compose it are far more important.

## National income, a measure of economic activity

It should be noted that National Income statistics relate mostly to economic activities, i.e. activities pursued for money. Work done for oneself or voluntarily for others does not normally enter into National Income, not because it is irrelevant, but because it cannot be measured. For example, if a man pays to have his house painted, the value of the work is included; if he does it himself, it is not included. The statistical difficulties in estimating the value of unpaid work and services would be tremendous. The principal item would, of course, be unpaid domestic services. This is a good example of the problem that often arises in statistics, whether to choose a second-best concept for which good data are available or a better concept for which satisfactory data do not exist. Also excluded are payments which do not correspond to goods or services received, i.e. transfer payments (see page 287).

In addition to money incomes, however, National Income includes a certain amount of 'imputed' income, notably the 'rent' of owner-occupied houses and such forms of income in kind as can be identified, e.g. the equivalent of the food and clothing of the Armed Forces. There is a good economic argument for including the services of all kinds of durable goods, but it would be impracticable to attempt to value the services of such articles as furniture and motor-cars. The inclusion of imputed rent removes anomalies that might otherwise arise, e.g. if the National Income were reduced as a result of tenants buying the houses they already occupied.

## Capital consumption

As already mentioned, National Income is taken to mean net national income or product after allowing for depreciation, or capital consumption, i.e. the annual use of capital through wear and tear. Until 1953, depreciation was measured by the total of depreciation allowances for purposes of taxation, but this was far from ideal. For a time the attempt to measure real depreciation was abandoned, and from 1953 to 1955 no estimates of net national income were published. However, estimates of capital consumption at current prices are now a regular feature of the Blue Books. These estimates rest on somewhat arbitrary assumptions about the life of assets, the way in which they depreciate with time, and the adjustments required to express that depreciation in terms of current values.

It will be convenient to discuss Gross National Product (commonly abbreviated to GNP) rather than National Income, except in general terms. Each of the three aspects of Gross National Product will now be discussed in detail, beginning with that of production.

## The approach via production

*Gross Domestic Product at factor cost* is the total income from economic activity in the United Kingdom, or the value of all work done and services rendered valued at factor cost, i.e. at the prices paid to the various 'factors of production'—labour, capital, land etc. It is gross in the sense that no provision is made for depreciation. The main contribution comes from industry and resembles the net output of industry as defined for the Census of Production, except that it omits purchased services such as advertising and insurance. An adjustment is also made for 'stock appreciation', discussed below.

With the addition of net income from abroad, Gross Domestic Product becomes *Gross National Product*. That is to say, while the Domestic Product is what the United Kingdom produces, the National Product is the total income of residents in this country from production anywhere in the world.

The constituent items of Gross Domestic and National Product are shown in Table 34. Table 1.11 of the 1976 Blue Book, from which most of these figures are taken, gives a run of eleven years, but for reasons of space only one year is given here. The figures are subject to revision in later Blue Books. Items 13, 14 and 15 require some explanation.

**Table 34** Gross National Product in 1975 by industry or other source

| Industry or other source | £ million |
|---|---|
| 1. Agriculture, forestry and fishing | 2,527 |
| 2. Mining and quarrying | 1,645 |
| 3. Manufacturing | 26,726 |
| 4. Construction | 6,411 |
| 5. Gas, electricity and water | 2,866 |
| 6. Transport | 5,753 |
| 7. Communication | 2,809 |
| 8. Distributive trades | 9,159 |
| 9. Insurance, banking, finance and business services | 7,727 |
| 10. Public administration and defence | 7,107 |
| 11. Public health and educational services | 7,154 |
| 12. Other services | 10,430 |
| 13. Ownership of dwellings | 5,535 |
| 14. Adjustment for financial services | −3,623 |
| 15. Residual error | 920 |
| 16. Gross Domestic Product at factor cost | 93,146 |
| 17. Net property income from abroad | 949 |
| 18. Gross National Product | 94,095 |

(*a*) Ownership of dwellings (Item 13) includes imputed income from owner-occupied houses etc.

(*b*) The contribution of Insurance, Banking and Finance to the GNP is measured, before deducting financial companies' net receipts of interest, which are duplicated in the other items, necessitating an adjustment for financial services' (Item 14).

(*c*) *Residual error* (Item 15). As already mentioned, the estimates of Gross National Product from income and from expenditure are built up from largely independent data, and the differences are shown as 'residual error'. Purely for convenience of presentation the estimates from expenditure are taken as correct, and residual error imputed to those from incomes, but the latter estimates are not necessarily less accurate than the former.

## The approach via incomes

The Gross National Product can also be presented to show the wages, salaries, rents, profits etc., comprising the net output of industry and the value of other work and services. The bulk of this total is personal

**Table 35**   Gross National Product in 1975 by type of share

(Adapted from the 1976 Blue Book, Tables 1.1 and 4.1)

| Share in the gross national product | £ million |
|---|---|
| 1.  Wages and salaries | 59,292 |
| 2.  Pay in cash and kind of HM Forces | 1,296 |
| 3.  Employers' contributions—National Insurance etc. | 4,043 |
| 4.  Employers' contributions—Other | 3,550 |
| 5.  Total income from employment | 68,181 |
| 6.  Professional persons | 1,322 |
| 7.  Farmers | 1,767 |
| 8.  Other sole traders and partnerships | 5,616 |
| 9.  Total income from self-employment | 8,705 |
| 10.  Gross trading profits of companies | 10,387 |
| 11.  Gross trading surplus of public corporations | 2,898 |
| 12.  Gross trading surplus of other enterprises | 114 |
| 13.  Rent | 7,144 |
| 14.  *less* Stock appreciation | −5,203 |
| 15.  Residual error | 920 |
| 16.  Gross Domestic Product at factor cost | 93,146 |
| 17.  Net property income from abroad | 949 |
| 18.  Gross National Product | 94,095 |

income, but as Table 35 shows, profits of companies make a large contribution.

Items 6 to 13 are all gross before allowing for depreciation. Several items of this table need comments and explanations:

(*a*) Wages and salaries (Item 1) are gross before deductions for income tax, National Insurance contributions etc.

(*b*) Employers' contributions—other (Item 4) includes employers' contributions to pension funds and, in the case of unfunded schemes, pensions paid to retired employees, also redundancy and compensation payments.

(*c*)  Rent (Item 13) includes imputed rent as before.

(*d*) Changes in the value of stocks are divided into (i) the value of the physical increase or decrease in stocks, (ii) the change in value arising from increases in prices at which stocks are valued. The latter part, called *Stock Appreciation* (Item 14) is treated as a capital gain and must be deducted from total income. It is estimated by applying price indices to as many categories of stock as data permit, and is admitted to be highly uncertain.

*Personal income* generally means the income of an individual, or of a married couple, as assessed for income tax, i.e. it includes all earnings, dividends, family allowance etc., but excludes expenses deducted for tax purposes. *Total personal income* includes not only the total of personal incomes as defined above, but also the (as yet) undistributed profits of life assurance and pension funds, imputed income from house-ownership, and various other items such as accrued interest on National Savings Certificates. It also includes 'current grants to persons', i.e. National Insurance benefits, family allowances and other grants from public authorities, and interest on the National Debt.

The last group are often called *transfer payments*, because they are transfers of income from one group of persons, e.g. tax-payers or rate-payers, to another group, not payments for goods produced or services rendered (unless maternity benefits and family allowances can be so described). It should be noted that some transfer payments, notably school milk and welfare foods, are not income as generally understood, and are not regarded as income for purposes of taxation, while others, such as pensions and family allowances, are subject to tax. None of them, however, is directly related to production. Furthermore, if all incomes are added together, these payments are counted twice, once in the tax-payer's income and once in that of the recipient. Total personal income therefore contains an element of duplication, and if the Gross National Product is derived from it, transfer payments must be deducted.

## The approach via expenditure

The goods and services made available for use during any given year include both domestic product and imports, and they can be either consumed (or bought for consumption), saved in the form of capital investment or addition to stocks, or exported. Table 36 shows how this brief analysis can be used to derive the Gross National Product from figures of expenditure, savings and foreign trade.

(*a*) *Consumers' expenditure* (Item 1) is expenditure on consumers' goods and services by persons and non-profit-making bodies, including the value of income in kind. In other words, it is not simply the amount of money spent, but the market value of the goods and services consumed. In the case of hire-purchase transactions, however, the whole of the purchase price is included.

(*b*) *Public authorities' expenditure* (Item 2) is current expenditure

**Table 36**   Gross National Product in 1975 by type of expenditure

(Adapted from the 1976 Blue Book, Table 1.1)

| *Expenditure generating gross national product* | *£ million* |
|---|---|
| 1. Consumers' expenditure | 63,373 |
| 2. Public authorities' current expenditure on goods and services | 22,907 |
| 3. Gross domestic fixed capital formation | 20,510 |
| 4. Value of physical increase in stocks and work in progress | −1,349 |
| 5. Total domestic expenditure at market prices | 105,441 |
| 6. Exports of goods and services | 26,093 |
| 7. *less* Imports of goods and services | −28,248 |
| 8. *less* Taxes on expenditure | −14,046 |
| 9. Subsidies | 3,906 |
| 10. Gross Domestic Product at factor cost | 93,146 |
| 11. Net property income from abroad | 949 |
| 12. Gross National Product | 94,095 |

constituting a direct demand for goods and services, and excludes transfer payments and debt interest, capital expenditure, loans etc.

(*c*) *Fixed capital formation* (Item 3) is expenditure on new buildings, plant and machinery etc., whether for addition to, or replacement of, existing assets, and excluding expenditure on repairs and maintenance, which is treated as a current cost of production.

(*d*) *The value of the physical increase in stocks and work in progress* (Item 4) is the increase in the value of stocks less stock appreciation and, like the latter, is subject to considerable error. In this table, it is negative because stocks were reduced in 1975.

(*e*) *Total domestic expenditure at market prices* (Item 5) contains all Customs and Excise duty, VAT etc. included in the market price of the goods and services bought, and these items must be deducted to arrive at total domestic expenditure at factor cost. Similarly, any subsidies reducing the market price below factor cost must be added back.

The Blue Books give several tables showing details of consumers' expenditure at current prices and also at fixed prices. Thus, in the 1976 Blue Book, expenditure is revalued at 1970 prices (see page 254).

The only way to estimate personal saving is to deduct personal expenditure from *personal disposable income*, i.e. personal income after tax and National Insurance contributions. As residual figures, these estimates are subject to very large errors and subsequent corrections, which provoked *The Economist* to say, some years ago, 'A separate chapter headed "Where we went wrong last year" should be an annual feature of the White Paper'.

## Social accounts

Most of the tables in the Blue Book are in the form of accounts, called *social accounts*, in which payments by, and receipts of, a group or *sector* of the economy are so defined that their totals must balance. It is customary to base these accounts on the following sectors:

The personal sector (including unincorporated business concerns).
Companies (private corporate trading bodies).
Public corporations (e.g. nationalized industries).
Central Government including National Insurance funds.
Local authorities.
The rest of the world (a 'sector' representing non-residents' transactions with the United Kingdom).

It may be observed that other countries do not necessarily divide their economies into the same sectors.

It is convenient also to distinguish between three types of economic activity, viz. production, consumption, and addition to wealth by capital formation. Each of these could in theory have its corresponding type of account, an Operating Account for production, an Income and Expenditure Account or Appropriation Account for consumption, and a Capital Account for addition to wealth, but in practice many of these accounts would not be appropriate. Table 37 shows the Combined Operating Account for Public Corporations for 1975.

A distinction can also be drawn between types of transaction, e.g. payments for goods and services, and transfer payments.

## 'Real' national income

The chief difficulty in studying movements in the National Income or its components is that they are expressed in terms of money, the 'value', or purchasing power, of which is falling from year to year. Attempts have therefore been made to measure the 'real' National Income or product, i.e. the volume of goods produced and services rendered. There are two intelligible ways of doing this. One is to extend the methods and information used in compiling the Index of Industrial Production to cover agriculture, services, and other forms of economic activity, giving the results in the form of index numbers.

**Table 37** Combined Operating Account for Public Corporations, 1975
(Adapted from 1976 Blue Book, Table 6.1)

| Receipts | £ million | Payments | £ million |
|---|---|---|---|
| Sales— | | | |
| (a) Revenue Sales | 16,373 | Wages, salaries etc. | 7,561 |
| (b) Sales to own capital | | Purchases of goods and ser- | |
| account | 725 | vices | 8,215 |
| Subsidies | 1,213 | *less* Increase in value of | |
| | | stocks and work in pro- | |
| | | gress | −838 |
| | | Taxes on expenditure: | |
| | | Rates | 298 |
| | | Motor vehicles and catering | |
| | | licences etc. | 12 |
| | | ECSC levies | 8 |
| | | Trading surplus and rent | |
| | | before providing for de- | |
| | | preciation and stock ap- | |
| | | preciation | 3,055 |
| Total | 18,311 | Total | 18,311 |

The other is to revalue the goods and services comprising the real product in terms of prices and values current in a particular year.

Both methods are used in the Blue Books. For example, the 1976 Blue Book contains one table entitled 'Expenditure and Output at 1970 prices' and another entitled 'Index numbers of output at constant factor cost (1970 = 100)'. The first table is derived from estimates of expenditure on goods and services, valued at 1970 prices; the second is based on the Index of Industrial Production and other statistics relating to volume of output.

Either method presents difficulties, particularly in dealing with services such as distribution and public administration. In the index number method, for example, insurance and medical services are measured by the number of beneficiaries served. The chief difficulty in the expenditure method is to find appropriate price indices for deflating current values. It is commonly assumed that the purchasing power of money is simply the reciprocal of the Index of Retail Prices, and Parliamentary questions and answers regarding the value of the £ are usually based on this assumption. In fact, however, this index only covers the field of retail transactions. For the remainder of the National Expenditure, e.g. public services and investment, reliable price indices do not exist. Moreover, the compilers stress that many estimates are extremely doubtful and that in many cases there are changes in quality that cannot be taken into account. It should be noted that the term 'real National Income' is never used in the Blue Books.

Assuming that the real National Income of a country can be measured with reasonable accuracy, a rough indication of the average standard of living at different periods can be obtained by dividing the real National Income by the population. It must be remembered, however, that habits of life change considerably over long periods, and the age-composition of the population should also be taken into account.

International comparisons by means of real National Income or product per head are even more difficult. Certainly there are official rates of exchange, but they do not always represent ratios of purchasing power. Comparison of British and American National Income per head will give different results before and after the devaluation of the pound in 1967. Two countries may differ in their treatment of transfer payments, the Armed Forces, domestic services, and many other matters. Moreover, in some countries much work may be done by the family, e.g. on the land, or by forced labour. Considerable progress has been made by the United Nations and the OECD towards uniformity of concepts, but differences between highly-developed and backward countries seem to present insoluble difficulties.

## Suggestions for further reading

Blue Books on *National Income and Expenditure*
*Economic Trends*
Studies in Official Statistics, No 13, *National Accounts Statistics: Sources and Methods*
Studies in Official Statistics, No 25, *The Measurement of Changes in Production*
Edey, H. C., Peacock, A. T. & Cooper, R. A. *National Income and Social Accounting*. London: Hutchinson
Stone, R. & G. *National Income and Expenditure*. Bowes and Bowes.

# Appendices

# Appendix 1 The use of computers for statistical work[1]

The value of using automatic equipment for analysing large amounts of statistical data has been realized for a long time; indeed, the original punched-card equipment was devised by Hollerith to cope with the American Census of 1888. The punched card he designed is precisely the card which is still used for input to computers. Both IBM and ICT marketed electro-mechanical equipment long before electronic computers were invented; IBM, in fact, were started in 1912.

The last ten years have seen a vast development in the use of computers, the major reason being the change in costs. Whereas most costs, especially wages, have increased with inflation, the costs of computer equipment have vastly decreased. A very closely related parallel is the drop in prices of electronic calculators. The effect is that mini-computers are currently available which are as powerful as the largest computers of only ten years ago, and at a fraction of the price.

## Parts of the computer

1. *Input* to a computer is still most likely to be in the form of punched cards (see p. 14); modern card readers are capable of reading 1400 cards per minute. Paper tape may sometimes be used instead. More specialized input devices such as document readers or magnetic ink character readers are unlikely to be used by the statistical department.

Alternatively, a terminal may be used at a remote point to input to the computer: this will usually be either a typewriter or a visual display unit (VDU). Such a device is very convenient, because it can be placed in one's own office, and also extremely useful for testing newly written programs. It is, however, comparatively slow and not likely to be used for inputting large amounts of data.

[1] Prepared by Frank Blewett, M.Sc., M.I.S.

2. *Processing* is carried out within the central processing unit (CPU) of the computer according to a programmed sequence of instructions. The program may have been specially written for this application, or may be a package program written by a computer company for a standard calculation, such as regression analysis. In either case, the program is likely to have been written in a 'high-level' language such as ALGOL, FORTRAN or BASIC. These languages look much like English (or at least algebra) and programs inside the computer would translate them into binary instructions.

3. *Backing Store.* This will be either magnetic tape or magnetic disc. Tape is much cheaper than disc, but can only be read sequentially (as with a tape recorder), whereas discs can be read randomly, i.e. any point may be accessed at any time. In the main, statistical files are likely to be kept on magnetic tape. However, where files set up for some other purpose are used for statistical analysis, they may well be kept permanently on disc. Examples are a personnel records file, or a file of customer records set up for an invoicing system.

4. *Output* is most likely to be printed on a line-printer. A terminal may be used for convenience, but (as with inputting) is relatively slow and is not likely to be used for large amounts of output.

## Use of the computer by a statistical department

The options available are:

1. *To buy time*, i.e. to run programs on a computer belonging to someone else, probably a service bureau. The bureau would also arrange for input data to be punched on to cards.

This option is likely to be preferred where the statistical department only occasionally needs a computer, and where the organization does not own a computer for other purposes.

2. *Time sharing.* Here the user either rents or buys a terminal and accesses a remote computer to run his program. The computer may be a large central computer belonging to one's own company, or a bureau machine, where one pays for whatever time is used. The access may be by telephone or by a private line. Such terminals are now very commonly used by statisticians.

3. *To buy their own computers.* A statistical department is unlikely to justify a large computer installation purely for its own work. On the other hand, a mini-computer may well be cheaper to use than a terminal, given a large enough workload.

## Costs of using a computer

The main cost of using a computer is likely to be not the cost of the actual computer time, but the associated cost and inconvenience of finding a suitable package program (or possibly writing a program) and of preparing data for the computer. The data may need to be coded on to special data sheets and then punched on to cards; often, if a questionnaire is designed well, it may be punched directly by a good punch operator.

## Advantages of using a computer

The advantages can be summarized as greater speed, accuracy and ability to generate effortlessly output in whatever form required.

A computer is not likely to be worthwhile where the only statistics needed can be obtained by adding up columns of figures. It will be extremely valuable where one needs cross-tabulations on large amounts of data, such as one may well require in analysing the results of a survey. Even with small amounts of data, a computer is likely to be justified if the calculations are at all complex, such as for regression analysis or analysis of variance. One must, however, guard against the temptation to generate unnecessary tabulations, otherwise one is likely to be buried under an avalanche of paper.

## Suggestions for further reading

Clark, J. O. E. *Computers at Work*. London: Hamlyn Paperbacks.

Hollingdale, S. H. & Tootill, G. C. *Electronic Computers*. London: Pelican Books.

Kemeny, J. G. & Kurtz, T. E. *BASIC Programming*. London & New York: John Wiley.

McCracken, D. D. *A Guide to ALGOL Programming*. London & New York: John Wiley.

McCracken, D. D. *A Guide to COBOL Programming*. London & New York: John Wiley.

McCracken, D. D. *A Guide to FORTRAN IV Programming*. London & New York: John Wiley.

Murdick, R. G. & Ross, J. E. *Information Systems for Modern Management*. London: Prentice-Hall.

# Appendix 2 Suggestions for general reading

Many references and suggestions for further reading have already been given, particularly in the later chapters, but a little more general guidance may be helpful. There are many (far too many) books on statistics, and the author has limited himself to a few with which he is acquainted and which he considers suitable for readers of this work.

Some readers may find this book a little concise and may wish to read other books covering similar ground but in a more discursive manner. Two excellent books for this purpose are Croxton, Cowden and Klein's *Applied General Statistics* (Pitman), and Wallis and Robert's *Statistics: A New Approach* (Methuen). These are both large and fairly expensive books, going somewhat farther than this book into the theory of statistics, but explaining everything very clearly and in great detail. Croxton, Cowden and Klein is particularly useful for its chapters on time series and index numbers. Wallis and Roberts is valuable for its hundreds of examples and exercises.

A smaller and remarkably cheap book, of more general interest that its title would suggest, is Bradford Hill's *Principles of Medical Statistics* (The Lancet Ltd). In a very simple and lucid manner it deals with the collection and presentation of data, averages and dispersion, the elements of sampling and statistical tests, and the principles of experimentation, carefully pointing out the difficulties and fallacies encountered in these operations. There are also excellent chapters on standardized death rates and other indices, and life tables.

Ilersic's *Statistics* (H.F.L. (Publishers) Ltd) covers much the same ground as this book but is aimed at the less mathematical student. It contains excellent chapters on sampling methods and surveys, and vital statistics (including a detailed account of the construction of life tables).

The student wishing to pursue the mathematical theory underlying statistical techniques will find an admirably clear exposition of the subject in Hoel's *Elementary Statistics* and in his more advanced *Introduction to Mathematical Statistics*, both published by Chapman and Hall.

# Answers to exercises

For reasons of space, and also because there is often no single correct answer, no answers are given to exercises involving charts, tables or general discussion. (Even with those given below, there may sometimes be differences of opinion.) On the other hand, hints for solution and comments on the questions themselves are given when required. Answers are rounded off as appropriate.

## Chapter 6

6.1   $3 \cdot 03$ per cent; $0 \cdot 182$ cubic feet.

6.2   $1 \cdot 4648 \pm 0 \cdot 0029$.

6.3   Estimate $4 \cdot 023$, minimum and maximum possible values $2 \cdot 935$ and $5 \cdot 278$.

6.4   It is not quite clear what is expected, short of doing nine multiplication sums (without a machine, in the examination!) and three additions. The estimated, minimum and maximum values for the total are 2,174,360 tons, 2,165,682 tons and 2,183,052 tons.

6.5   Actual errors are $-£383$, etc., but to readers of the rounded figures the possible errors will be $\pm £500$ on each item ($\pm £3000$ on the total if it has been obtained by adding the rounded figures).

6.6   $31 \cdot 15$ m.p.g.; $28 \cdot 3$ to $34 \cdot 6$ m.p.g.

6.7   (i) £5350;
      (ii) £8542.50; £7462.50;
      (iii) £3542.50; £2112.50;
      (iv) £3.52; £2.12.

## Chapter 8

8.1   Period is 6 years, from peaks and troughs. Take 6-point moving totals, then add pairs and divide by 12.

8.3    Seasonally corrected figures: January £1250, February £1222, March £1211.

8.3    Average seasonal movements: Spring, +10; Summer, −24; Autumn, −11; Winter, +25. (The question does not state the unit.) The first model fits well.

8.4    About 1,700,000 and 3,200,000. (Actual figures were 1,420,000 and 3,231,000. The alternative model would clearly be more suitable in this case. It would give 1,430,000 and 3,220,000.)

8.5    1st quarter, +12; 2nd quarter, −11; 3rd quarter, −46; 4th quarter, +45. Seasonally adjusted figure, 858. Yes.

8.6    Seasonal adjustments: add 3·1, 0·6, 0·3 and −4·0 respectively. Forecast for 4th quarter of 1973, about 19·5.

## Chapter 9

9.2    A poor question. There are too few employees to draw a reliable frequency curve, and the distribution is not symmetrical as the question implies.

9.5    Check your answers from the original data, remembering that '1' means 'between $\frac{1}{2}$ and $1\frac{1}{2}$'.

## Chapter 10

10.1    (*b*) The result will vary slightly according to your grouping. Compare it with the average of the original data.

10.2    93·84 lb per sq inch.

10.3    The data are somewhat peculiar: seven items of 168 and only one item between 152 and 168! Who's been cooking the records? Your answers will depend somewhat on your grouping, but compare them with the mean (160), median (168) and mode (168) of the original data.

10.4    Arithmetic means (in minutes): *A*, 15·76; *B*, 13·40. Medians: *A*, 13; *B*, 15. To explain this anomaly, try drawing cumulative frequency polygons.

10.5    28 mph; 19 per cent.

10.6    Mean 25·44 tons, median 25·0 tons, mode 20·9 tons, but the mode does not make sense with only 12 discrete items. If we assume that these are 12 random items from a large population, we can estimate the mode as 24·12 from the formula

$$\text{Mode} = \text{Mean} - 3\,(\text{Mean} - \text{Median})$$

But does it really mean anything?

10.7    (i) 126.

(ii) 110.

10.8    Arithmetic mean, 9·375; geometric mean, 8·17; harmonic mean, 6·99. (Some arithmetic for 36 minutes!)

## Chapter 11

11.1    12·7 weeks; 22·2 weeks.

11.2    Median 99, quartiles 90 and 108.

11.3    Approximately 4′ 10·9″, 5′ 0·0″, 5′ 0·9″, 5′ 1·6″, 5′ 2·3″, 5′ 3·0″, 5′ 3·7″, 5′ 4·5″, 5′ 5·7″. The decimal figures are uncertain.

11.4    Very nearly 6 lb, measured from either mean or median.

11.5    (a) 1949 or earlier: mean 2·73, SD 1·63; 1955–9: mean 2·51, SD 1·29.

11.6    (i) 54·2; (ii) 12·4, or with Sheppard's correction, 12·1.

11.7    Mean 82·080, standard deviation 3·673.

11.8    Coefficients of variation: English 47 per cent, Arithmetic 40 per cent.

## Chapter 12

12.1    (a) $\frac{81}{256}$, (b) $\frac{3}{8}$.

12.2    (a) $\frac{1}{12}$, (b) $\frac{1}{2}$, (c) $\frac{1}{4}$, (d) $\frac{1}{2}$.

12.3    81 per cent.

12.4    (a) 293, (b) 6, (c) 31.

12.5    (a) Mean 171·94 lb, standard deviation (with Sheppard's correction) 29·82 lb.

(b) 210 lb.

12.6    (a) 9·1 per cent.

(b) 2·9 per cent (new mean 26·75 lb).

12.7    (a) 13·4 per cent.

(b) (i) 0·0743, (ii) 0·0758.

12.8    (a) (i) 0·1937, (ii) 0·1840.

(b) (i) 0·9298, (ii) 0·9198.

This question illustrates the inadequacy of the Poisson approximation when $p$ is as high as 0·1.

12.9    (b) 0·942 (Poisson approximation gives 0·938).

(c) 10.

12.10   (a) $e^{-4}$.

(b) $\frac{8}{3}e^{-1}$.

(c) $6e^{-6}$.

## Chapter 13

13.2 Highly significant; the campaign was successful (but what a laborious question without a machine!).

13.3 Primer $A$ is almost certainly superior ($P < 0.01$).

13.4 ($41 \pm 3.0$) per cent, or 38 per cent and 44 per cent.
Difference of 11 per cent is highly significant.

13.5 (*a*) 34 per cent.
(*b*) ($34 \pm 4.15$) per cent, or say 30 per cent and 38 per cent.
(*c*) About 2150.

13.6 (*a*) 9500.
(*b*) Stratify to remove bias.

13.7 The first part is stupid. The seedsman expects 40 seeds to germinate and 45 do. Of course he is justified.
The stock is almost certainly sub-standard. The two results are not consistent.

13.8 0.73 to 3.27. No evidence of change: difference between sample means is 1, standard error of difference is 0.95.

13.9 7, using a double-sided test.
3 hr 12 min; 3 hr 2 min and 3 hr 22 min.
Yes: the confidence interval does not contain the old time.

13.10 Yes: $t = 2.32$ for 9 df, $P < 0.05$.

13.11 Yes: there are 9 positive differences and one negative one, giving $P = 11/512$. Using the $t$-test, $t = 3.57$ for 9 df, $P < 0.01$.

## Chapter 14

14.2 (*a*) Any estimate between 20 and 21 knots is reasonable.
(*b*) Probably about 25 knots, but the relationship is not linear and the shape of the curve outside the range 6000–30,000 tons is very uncertain.

14.3 $y = 0.00217 (x - 45) + 0.35$.
Prediction for 100°C, 0.47, assuming that a linear trend is still valid at that temperature.

14.4 0.8. But fancy only doing this for five pairs of observations!

14.5 (*a*) 0.88.

14.6 $y = 8.24x + 5.43$; 13.67; 23.56; 26.03.

14.7 (*a*) 13.8 mn, (*b*) 13.6 mn.

14.8 Spearman's rank correlation coefficient is 0.82, but it is impossible to assess its significance with only seven items.

14.9    Spearman's rank correlation coefficient is 0·83, which is almost certainly significant. The laboratory test is a good guide.

## Chapter 15

15.1    $\chi^2 = 6·78$ for 5 degrees of freedom. Not significant.
15.2    $\chi^2 = 2·99$ for 2 degrees of freedom. Not significant.
15.3    $\chi^2 = 24·6$ for 6 df. Highly significant.
15.4    $\chi^2 = 30$ for 6 df. (No great precision required here.) Highly significant.
15.5    $\chi^2 = 5·22$ for 1 df. Probably significant.
15.6    Poisson with parameter 0·71. Group '2 or more'. $\chi^2 = 1·76$ for 1 df. Not significant.

## Chapter 16

16.1    1·4 per cent.
16.3    (*b*) 2·28 per cent.
16.7    Assume particles randomly distributed and use Poisson distribution with parameter 1.
        Yes: actual and expected frequencies agree well.
16.8    Process out of control, i.e. quality is variable.
16.9    (*a*) 3·4 per cent.
        (*b*) Inner limit 9·4, outer limit 12·1.
        (*c*) No. The mean proportion would be the same for smaller samples (presumably 'batches' is a mistake for 'samples'), but the standard error of the proportion would be increased.
16.10   Process SD is given, viz. 4 g, so the standard error of a sample mean is $4/\sqrt{4} = 2$ g. The mean range of a sample of 4 is $4 \times 2·06 = 8·24$ g. 'Confidence limits' is an obvious mistake for 'control limits'. Warning limits for the mean are

$$40 \pm (2 \times 1·96) = 36·1 \text{ g and } 43·9 \text{ g.}$$

Action limits are

$$40 \pm (2 \times 3·09) = 33·8 \text{ g and } 46·2 \text{ g.}$$

From tables, e.g. the IoS tables, warning and action limits for the range are respectively

$$8·24 \times 1·93 = 15·9 \text{ g,}$$
$$\text{and} \quad 8·24 \times 2·58 = 21·2 \text{ g.}$$

## Chapter 17

17.1  1971: 109·0; 1972: 126·5.

17.2  (a) 24, (b) 147·9, (c) 104 if the figures in the last column are taken as exact. Since they have been rounded, however, (c) could be 103, 105 or even 106.

17.3  Your answer will probably be Laspeyres, 123, or Paasche, 118.

17.4  Both Paasche and Laspeyres indices are 225·3 to one place of decimals. Careful examination of the data will show why.

17.5  (i) 103·8, (ii) 93·4.

17.6  (a) Divide Col (1) by Col (3).

17.7  Possibilities include value at 1962, 1963 or 1964 prices, simple arithmetic mean of production relatives, etc.

17.8  A difficult problem needing careful thought. If $c_0$ and $c_1$ denote 1970 and 1976 quantities of cheese and $\Sigma p_0 q_0$ etc. denote values of the other items,

$$\Sigma p_1 q_0 + 0{\cdot}7c_0 = 1{\cdot}25 \,(\Sigma p_0 q_0 + 0{\cdot}5c_0)$$

$$\text{and } \frac{0{\cdot}5c_0}{6} = \frac{\Sigma p_0 q_0}{94} = \frac{\Sigma p_1 q_1 + 0{\cdot}7c_0}{1{\cdot}25} = \text{etc.,}$$

which leads to

$$\frac{\Sigma p_1 q_0 + 0{\cdot}6c_0}{\Sigma p_0 q_0 + 0{\cdot}5c_0} \times 100 = 123{\cdot}8.$$

Similarly, the answer to the second question is 124·1. The fact that expenditure on cheese was 4·2 per cent of the total in 1976 is irrelevant to the first question.

## Chapter 19

19.1  (b) (i) 14 per 1000.
   (ii) 11 per 1000.

19.2  The answer will depend on the standard population chosen, but groups 'under 20' and '60 and over' must be omitted, as they have no means of comparison.

19.3  Actual, 11·60 days; expected 10·99 days.

19.4  Crude death rate, 13·2 per 1000; standardized death rate, 13·5 per 1000.

19.5  Remember that the figures of deaths cover a period of five years. The SMR is 91.

19.6  6787 school places needed; 31,024 jobs required; 71 per cent.

19.7    The missing figures are as follows:

| Age $x$ | $l_x$ | $d_x$ | $p_x$ | $q_x$ |
|---|---|---|---|---|
| 20 |  | 43 |  | 0·00044 |
| 21 | 97,293 | 44 | 0·99955 |  |
| 22 | 97,249 | 46 | 0·99953 |  |
| 23 | 97,203 |  | 0·99951 | 0·00049 |
| 24 | 97,155 |  | 0·99949 | 0·00051 |

# Index

Absence from work, 13
Acceptance sampling, 201–3
Accuracy and error, 39–44
Age-specific rates, 239
Aggregative method, 214, 217–19
Agricultural Prices, Index Numbers
    of, 254
Agricultural production, 265
*Agricultural Statistics*, 250, 265
Algol, 296
*Annual Abstract of Statistics,* 233
*Annual Statement of the Overseas
    Trade of the United Kingdom*,
    278
Approximation, 40–1
Area Comparability Factor, 242
Arithmetic mean, 93–6,
    compared with median and mode,
      99–101
    short-cut methods for, 95–6
    weighted, 96–7, 215–16
Array, 108
Assignable causes of Variation, 195
Attributes, 7
Averages, 93–103
    moving, 63–66
    (*See also* Arithmetic Mean, etc.)

Backing store, 296
Balance of payments, 277, 280–1

Balance for official financing, 280
Balancing item, 280
Bar charts, 48–51
Base date, 210
Base period, 210
Basic, 296
Basic records, 11–13
Bias, 40
    in surveys, 22–4
Binomial distribution, 130–1, 187–9
Birth rate, 238–9
*British Labour Statistics Yearbook*,
    257
Broken scales, 54
Budgets, household, 252
*Business Monitors*, 234, 266, 271,
    278
Business Statistics Office, 232–4

Capital consumption, 284
    expenditure, 273–5
    formation, 273
Caption (of table), 30
Card records, 12
Census of Agriculture, 265
    of Distribution, 232, 271–3
    of Employment, 257–8
    of Population, 231, 236–8, 244–5
    of Production, 231, 265–6
Censuses and surveys, 21–7

Central processing unit, 296
Central Register of Businesses, 232
Central Statistical Office, 232–4
Chain base method, 212, 222
Chance causes of variation, 195
Charts, 47–58
   (*See also* Control Charts, etc.)
Chi-squared, 182–92
Circular test, 224–5
Class, 79
   frequency, 79
   interval, 79
   limits, 81–2
Classification, 5–7
   changes in, 5, 35
   International Statistical, of
     Diseases, Injuries and Causes
     of Death, 18, 241
   Standard Industrial, 232, 258–9
Clerical Salaries Analysis, 262–3
Cluster sampling, 26
Coding, 13–14
Coefficient of correlation, 173–5
Coefficient of rank correlation,
   175–7
Coefficient of regression, 169
Coefficient of skewness, 119
Coefficient of variation, 117–18
Cohort analysis, 239
Collection of data, 5
Comparisons—
   international, 269–70, 291
   paired, 157–9
   validity of, 34–5
Computers, 295–7
Conditional probability, 126
Confidence interval, 147
Confidence level, 148
Confidence limits, 147
Consistency tests (index numbers),
   223–5

Consumers' expenditure, 254, 287
Consumers' goods, 264
Consumer's risk, 201
Contingency tables, 184–6, 258–9
Continuity, breaks in, 35, 258–9
Continuity, correction for, 136
Continuous variable, 78
Control (of processes), 195
Control charts, 195–201
   for attributes, 199
   for variables, 195–9
Control limits, 196–9
Correlation, 166
   coefficient of, 173–5
   direct and inverse, 166
   rank, 175–7
Cost of Living Index Number,
   251–2
Cumulative frequency curves, 88–90
Cumulative frequency distributions,
   87–90
Cumulative sum charts, 200
Current balance (of trade), 280
Curve fitting, 63
Curve of Errors, 133
Curve of regression, 168
Customs and Excise Tariff, 277
Cyclical fluctuations, 62, 25

Data—
   collection of, 5
   internal and external, 5
   primary and secondary, 5
Date base, 211
Death, causes of, 236, 241
Death rates, 9–10, 239–43
   age-specific, 239
   crude, 239
   equivalent annual, 9, 239
   standardised, 241–3
Deciles, 107

Defective, proportion, 199
Definition—
  differences of, 6
  importance of precise, 4, 35, 258
Deflated value, 267
Degrees of freedom, 150–1
Demographic statistics, 236–46
*Department of Employment Gazette,*
  233, 257
Depreciation, 284
Derivative data, 8–9
Descriptive statistics, 1
Diagrams, 47
Differences, algebraic and absolute,
  8
Discrete variable, 78
Dispersion, 93, 107–18
  absolute and relative, 117
  various measures of, 117
  (See also Mean Deviation, etc.)
Distribution—
  Census of, 232, 271–3
  fitting, 186–91
  frequency, 78–90
  probability, 127–9
  sampling, 144–5
Distribution curves, 89–90
Distribution-free tests, 159–60
Distributions, combining two or
  more, 116
Double sampling, 202
Double-sided tests, 153

Earnings, 260–2
Earnings and hours of wage-earners,
  inquiries into, 261–2
*Economic Trends,* 233
Electoral Register, 24, 26
Electronic computers, 295–7
Employment, 257–9
  Census of, 257–8

*English Life Table No. 12.,* 240
Error—
accuracy and, 39–44
  biased and unbiased, 40
  kinds of, 39–40
  possible, 43
  residual, 285
  sampling, 39
  standard, 145
Errors—
  absolute and relative, 42
  combination of, 42–4
  Curve of, 133
Errors in Census data, 237
Errors of measurement, 39
Errors of origin, inadequacy and
  manipulation, 39–40
Estimates, comparison of two,
  153–7
Expectation (expected value),
  129–130
Expectation of life, 240–1
Expenditure, National, 283, 287–9
Exports, 277–80
  indices of unit value and volume
  of, 278–80
Exposed to risk, 9, 19, 239

Factor cost, 266, 284
Factor-reversal test, 223–4
Factorial experiments, 205
Family Expenditure Survey, 253
Fertility, 238–9
Field staff, 23
Finite population correction factor,
  147
Fisher, Irving, 222
Fisher's 'Ideal' Formula, 222, 224–6
Fitting a distribution to frequency
  data, 186–91
Fixed-base methods, 222

Forecasting, 73
Forecasts, 2, 16, 244
Fortran, 296
Frequency, 78
  curves, 85–7
  diagrams, 47, 82–90
  distributions, 78–90
  polygons, 85

Gallon, Imperial and U.S., 8
Galton, Sir Francis, 165
Gate (in tabulating), 34
Gauss, 133
Gaussian distribution (*See* Normal)
General Household Survey, 233, 246
General Register Office, 232, 236
General Replacement Rates, 239
Geometric mean, 101–2, 216–17, 250
Goodness-of-fit tests, 187
Gosset, W. S. ('*Student*'), 150
Government Social Survey, 232
Graphs, 47
Gross Domestic Product, 284
Gross margin, 272
Gross National Product, 284–8
Gross output, 266
Gross Reproduction Rate, 239
Group (class), 79
*Guide to Official Statistics*, 234

Harmonic mean, 102–3
Hire-purchase statistics, 273
*Historical Abstract of Labour Statistics,* 257
Histograms, 82–5
Homogeneity, 7
Hourly Rates of Wages, Indices of, 260–1
Hours of Work, 260–2

Household budgets, 252
Household Expenditure Enquiry, 253
Hump-backed distributions, 86
Hundredweight, U.S. and Canadian, 8
Hybrid formul.formulae, 221–2

'Ideal' Formula, 222, 224–6
Imports, 277–80
  indices of unit value and volume of, 278–80
Income—
  imputed, 283, 285
  National, 282–91
  personal, 287
Index Death Rate, 242
Index numbers, 210–26
  changes in quality of items for, 212
  choice of items for, 211–12
  criteria of good, 225
  methods of calculation of, 214–22
  progressive changes in, 213–14
  tests of consistency for, 223–5
  weighting of, 210–11
Indicators, 62, 254, 265
Industrial applications of statistical methods, 195–205
Industrial production, 264–8
  Index of, 267–8
Infant mortality rate, 239–40
Input (computer), 295
Input-output tables, 268–9
Inquiry—
  full-scale or sample, 21–2
  mail, 23–4
Institute of Administrative Management, 262
Institute of Statisticians, 21
Internal trade, 271–6

International Passenger Survey, 244, 281
International Statistical Classification of Diseases, Injuries and Causes of Death, 18, 241
Interpolation, 98, 109–10
Interviewer bias, 23–4
Interviews, 23–4
Inverse correlation, 166
Invisibles, 280–1

J-Shaped distributions, 87

Labour statistics, 257–63
Labour turnover, 18–19, 259–60
Languages, computer, 296
Laspeyres, 218–22, 224–6
Law of Statistical Regularity, 21–2
Least Squares, method of, 168–70
Life tables, 240
Line charts, 51–4
Logarithmic charts, 56–8
Loose-leaf records, 12
Lord's test, 159

Macmillan, Harold, 232
Mail inquiries, 23
Market research, 23
Mathematical statistics, 1
Mean (*See* Arithmetic mean, etc.)
Mean deviation, 111
Median, 97–8
Method of Least Squares, 168–70
Migration, 244
Mode, 99
Model, 70
Modulus, 8
Monte Carlo methods, 204
*Monthly Digest of Statistics,* 35, 233
Mortality rates (*See* Death rates)
Moving annual total, 66

Moving average, 63–6
Multi-stage sampling, 26

National expenditure, 283, 287–9
National income, 282–91
  international comparisons of, 291
  real, 289–91
  threefold nature of, 282–3
*National Income and Expenditure* (Blue Book), 233, 254, 282
*National Institute Economic Review,* 234
National Insurance cards, 257–8
Net output, 266
Net Reproduction Rate, 239
New Earnings Survey, 261
Nitrogen content of fertilisers, 8
Non-paramatric tests, 159
Non-response, 22–3
Nonsense correlation, 175
Normal curve, 131–6
Normal deviate, 133
Normal distribution, 131–6, 144–5, 147, 190–1
Normal probability paper, 135
Null hypothesis, 152, 160

Occupational mortality, 241
Office of Population Censuses and Surveys, 232, 236
Official statistics, 5, 231–4
Ogive, 89, 109–10
O.P.C.S. Monitors, 245
Operational research, 203–4
Output (Computer), 296
Output, gross and net, 266
Output per head, 269–70
Overseas trade, 277–81
  *Overseas Trade Statistics,* 278

Paasche, 218–22, 224–6

Paired comparisons, 157–9
Parameters, 93
Payment in kind, 260
Payments, Balance of, 277, 280–1
Percentages, 9
Percentiles, 107–8
Personal income, 287–8
Personnel statistics, 18–19
Pie charts, 48
Pilot survey, 24
Pitfalls, 5–6
Poisson distribution, 136–8, 189–90
Population—
    age and sex structure of, 10
    Census of, 231, 236–8
    projections of, 244
    working, 258
Post-enumeration survey, 237
Precision, 40
    spurious, 42
Price relatives, 210, 214–17, 250–1, 254
Prices, 250–6 (*See also* Retail Prices, Wholesale Prices)
Primary data, 5
Probability, 124–38
    addition law of, 125
    conditional, 126
    multiplication law of, 125
Process average and variability, 195
Processing (computer), 296
Producers' goods, 264
Producer's risk, 201
Product—
    Gross Domestic, 284
    Gross National, 284–8
Production, 264–70
    Agricultural, 265
    Census of, 231, 265–6
    Index of Industrial, 267–8
    Industrial, 264–8

Production planning, 204
Production statistics, 17
Productivity, 269–70
Projections of population, 244
Proportion, standard error of, 146–7
Public opinion polls, 27
Published statistics, 231–4
Punched card equipment, 14–15, 295
Punched cards, 14

Qualitative factors, 6–7
Quality control, 195
Quantiles, 111
Quantitative factors, 7
Quantity, index numbers, 219–21
Quantity, relatives, 219, 267
Quartile deviation, 111
Quartiles, 108–10
Quasi-random sampling, 26
Questionnaires, 24
Quota sampling, 27

Random deviations, 69
Random numbers, 25
Random sampling, 25
Range, 107
    control charts for, 196–8
Rank correlation, 175–7
Rates, 9–10
Ratios, 8–9
Records, basic, 11–13
Rectangular distribution, 187
Rectifying inspection, 203
*Regional Statistics,* 233
Registrar-General, 236
Registration of births, deaths and marriages, 236
Regression and correlation, 165–177
Regression coefficients, 169
Regression diagrams, 167–8

Regression equations, 168–172
Regression lines, 168–172
Relative error, 42
Relatives (*See* Price Relatives, Quantity Relatives)
Reproduction Rate—
  Gross, 239
  Net, 239
Residual error, 285
Residual variations, 62, 195
Response errors, 24
Retail Prices, Index of, 251–6
Retail Sales, Index Numbers of, 271–3
Returns, statistical, 12–13
Robust tests, 159
Rounding, 40–1
Routine statistics, 11–19

Salaries, 262–3
Sales forecasts, 16
Sales statistics, 15–16
Sample Census of 1966, 21, 237–238
Sample means, comparison of, 155–7
Sample proportions, comparison of, 154–5
Sampling, 21–7, 142–60
  acceptance, 201–3
  cluster, 26
  double, 202
  quota, 27
  random, 25
  sequential, 202
  single, 202
  stratified, 26
  systematic, 25–6
Sampling distributions, 144–5
Sampling errors, 39
Sampling frame, 24

Sampling in Census of Population, 238
Scales (graphs), 54–5
  broken, 54
Scatter diagrams, 48, 166–7
Seasonal factors, 71–3, 268
Seasonal variations, 62, 66–73
Seasonally corrected figures, 67–9, 273–4
Secondary data, 5
Sector of economy, 289
Semi-longarithmic charts, 56–8
Sequential sampling, 202
Sheppard's correction, 115–16
Sickness, 18, 244–5
Sign test, 159
Significance—
  level of, 152
  tests of, 152–160
Simulation, 203–4
Single sampling, 202
Single-sided tests, 153
Skew distributions, 86
Skewness, 93, 118–19
Smoothing, 63–6
Social accounts, 289
*Social trends*, 233, 245
Spearman's rank correlation coefficient, 175–7
Spurious precision, 42
Standard deviation, 112–17
  best estimate of, 149–50
Standard deviation of a sum, difference or mean, 142–4
Standard error, 145
  of mean 145, 150
  of median, 147
  of proportion, 146–7
  of standard deviation, 147
Standard Industrial Classification, 232, 258–9

Standard Regions, 232
Standardised rates, 241–3
  Morality Ratio, 243
Statistic, 144
Statistical data, 4–10
*Statistical News,* 233
Statistical Regularity, Law of, 21–2
Stock appreciation, 286
Stock control, 203–4
Stocks, 17, 203–4, 273–4
Stockturn, 272
Store (computer), 296
Strata charts, 55–6
Stratified sampling, 26
Stub (of table), 30
'Student', 150
*Studies in Official Statistics,* 233–4
Supply statistics, 17–18
Surface charts, 55–6
Surveys, 21–7
Systematic sampling, 25–6


Tables, 30–5
  hints on construction of, 34–5
  structure of, 30–3
Tabulation, mechanical, 237
t-distribution, 150–1
Terms of trade, 279–80
Tests of consistency, 223–5
Tests of significance, 152–160
Time-reversal test, 223
Time series, 62–75
  charts of, 51–4
Time sharing, 296
Title (of table), 34
Tons, long, short and metric, 8
Total personal income, 287
Trade—
  internal, 271–6
  overseas, 277–81

  terms of, 279–80
  visible, 280–1
*Trade and Industry,* 233, 278
Trade cycle, 62
Transfer payments, 287
Transport, 275–6
Trend, 62–3


Unbiased errors, 40
Unemployment, 259
Unemployment Insurance Acts, 5, 258
Unit value indices, 212
  of imports and exports, 278–9
Units (of measurement), 7–8, 34
  simple and composite, 7–8
  variables in different, 55
Universe, 145


Value added by manufacture, 266
Variability in observed data, 39
Variable (or variate), 7, 78
Variables in different units, 55
Variance, 112
Variation, coefficient of, 117–18
  in industrial processes, 197
Visible index, 12
Visible trade, 280–1
Visual display unit, 295
Volume indices, 219–21
  of imports and exports, 278–9


Wage rates, 260–1
  Index numbers of, 260–1
Weighted average (or mean), 96–7, 215–17, 219–20, 251, 260, 273
Weights (in index numbers), 210–11, 219–20, 225–6, 251, 253–4, 260, 267, 273

Weights base, 211
Wholesale price index numbers, 213, 250–1
Wholesale prices, 250

Working population, 258
Working sheets, 33–4

Z-charts, 73–5